Electrical Hazards and Accidents

Their Cause and Prevention

Electrical Hazards and Accidents
Their Cause and Prevention

Edited by

E.K. Greenwald
College of Engineering
University of Wisconsin-Madison

VAN NOSTRAND REINHOLD
_____ New York

Copyright © 1991 by Van Nostrand Reinhold

Library of Congress Catalog Card Number 91-23009
ISBN 0-442-23799-5

Manufactured in the United States of America

Published by Van Nostrand Reinhold
115 Fifth Avenue
New York, NY 10003

Chapman and Hall
2-6 Boundary Row
London, SE1 8HN, England

Thomas Nelson Australia
102 Dodds Street
South Melbourne 3205
Victoria, Australia

Nelson Canada
1120 Birchmount Road
Scarborough, Ontario M1K 5G4, Canada

16   15   14   13   12   11   10   9   8   7   6   5   4   3   2   1

Library of Congress Cataloging-in-Publication Data

Electrical hazards and accidents: their cause and prevention / edited
    by E.K. Greenwald
        p.   cm.
    Includes index.
    ISBN 0-442-23799-5
    1. Electric engineering—Safety measures.   2. Industrial safety.
I. Greenwald, E.K.
TK152.E49   1991
621.319′24′0289—dc20   91-23009
                                              CIP

Contributors

Theodore Bernstein, Ph.D., P.E., Professor Emeritus, College of Engineering, University of Wisconsin and private engineering consultant and lecturer on lightning and electrical fire investigation
Madison, Wisconsin

Gregory Bierals, President, Electrical Design Institute, Inc.; Private consultant, lecturer, and National Electrical Code® registered trademark of National Fire Protection Agency, Quincy, Massachusetts expert
Davie, Florida

Glenn Schmieg, Ph.D., Retired college professor; private consultant and lecturer on lightning phenomena and static electricity
Milwaukee, Wisconsin

Marty Martin, P.E., Private engineering consultant with Arnold and O'Sheridan, Inc. and lecturer on electrical design
Madison, Wisconsin

E.K. Greenwald, Ph.D., P.E., Associate Professor, College of Engineering, University of Wisconsin and private engineering consultant with Energy Analysis, Inc.
Madison, Wisconsin

Acknowledgment

I am grateful to Professor Neva Greenwald for her review and many valuable suggestions in the preparation of this book.

Contents

Preface

As a professor of Engineering Professional Development, I develop and teach continuing education short courses for the working engineer. Every day practicing engineers request my assistance in finding sources for information and educational programs on various topics in engineering.

Over the last five years, in response to many requests, I have developed and offered annually a series of courses focusing on electrical safety, which are taught in the United States and abroad. To teach these courses I assembled a group of instructors whose expertise is recognized nationwide. Course titles include: Fundamentals of Electrical Fire Investigation, Lightning Protection, Electrical System Design in Hazardous Locations, and Grounding of Electrical Distribution Systems.

A number of engineers have requested a general textbook or reference for specific information on the subject of electrical safety in the industrial workplace. To satisfy this need, I invited four of my instructors to be contributors and assist me in preparing a reference book on industrial electrical safety. This work is the product of our collaborative effort.

E. K. GREENWALD

Introduction

Potential hazards and problems associated with electrical systems and equipment utilized in a modern industrial environment are discussed herein. Alternative approaches and solutions to problems are presented that will help the plant engineer to make the industrial environment a safer place in which to work.

This book should be useful to electrical engineers, production engineers, maintenance personnel, and anyone else involved in the delivery of physical plant services. It will be of particular assistance to supervisory and engineering personnel concerned with plant safety or physical plant engineering.

Chapter 1 explains the fundamentals of electrical systems and components that one needs to comprehend the ideas presented in this book. Among the subjects discussed are electrical codes and standards, definitions and electrical terminology, single and three-phase systems, electrical components, basic grounding concepts, fuses, plugs, ground fault circuit interrupters (GFCIs), and the physics of the electric arc and heating effects.

Chapter 2 discusses the physiological basis of the human response to an electric shock. Major biological topics include the approximate electrical impedance of the human body, thresholds of shock perception, let-go currents, asphyxia, ventricular fibrillation, and respiratory arrest. Lightning, high and low voltage shocks, power line accidents, and safe voltage levels also are considered.

Chapter 3 explains the various factors involved in determining the correct wire size for a conductor in an electrical application as mandated in the National Electrical Code (NEC). Then the correction factors that are applicable in special situations, such as a high ambient temperature environment, are analyzed. Finally the procedures mandated within the NEC for sizing the feeder and branch circuit overcurrent protection devices are described.

Chapter 4 presents the NEC requirements and procedures for grounding an electrical distribution system. The various types of grounding electrodes allowed by the NEC for specific applications are discussed. The quality requirements for the grounding electrode as well as the factors that influence soil resistivity are described.

Chapter 5 delineates the requirements of the Occupational Safety and Health Administration (OSHA) with respect to electrical systems in the workplace, with suggestions for compliance. A historical review of the legislation is presented with a detailed listing of retroactive and mandatory provisions, including the criteria for acceptance of an electrical system by OSHA.

Chapter 6 reviews the requirements for electrical systems located in hazardous locations. The different divisions and classifications are defined and discussed. Other topics considered are potential ignition sources, ventilation requirements, surface temperature conditions, conduit threads, and sealing requirements. A detailed case study describing how to classify a hazardous area is presented.

Chapter 7 discusses the behavior of current-carrying conductors in a fire environment. Topics presented include insulation behavior, conductor melding temperatures, and the effects of nicks and broken strands. Data for steady state and transient heat transfer from current-carrying conductors are provided. Then methods for investigating the fire scene are presented, followed by several case histories of actual fires.

Chapter 8 explains the basics of lightning theory. Isocenauric levels are defined and discussed. Then a method for determining the probability of having lightning strike at a specific location is presented. Current lightning protection models are presented (rolling ball, cone), along with protection methods for mitigating the effects of a direct lightning strike. Voltage surges are defined with an explanation of their possible sources. The discussion of voltage surge protection is followed by a presentation of investigative techniques to use in suspected lightning accidents.

Chapter 9 presents potential and real problems concerning static electricity and their solutions in the industrial environment. A discussion of the fundamentals of electrical charge induction is followed by an explanation of the mechanisms for static charge ignition. A detailed procedure for analyzing potential static electricity problems is presented and several case histories illustrate actual problems solved with this procedure.

Chapter 10 discusses the design, construction,maintenance, and operational procedures required to establish and maintain a safe working environment around high voltage systems in an industrial plant.

After Chapter 1's presentation of fundamentals, the topics that follow are essentially unconnected and independent of one another. The reader may freely choose readings by chapter, according to subject matter or his or her informational needs, without having to read the entire book. The authors have assumed that the reader will have a general technical background, including some knowledge of industrial electrical systems and the National Electrical Code®.

® Registered Trademark of National Fire Protection Agency, Quincy, Massachusetts.

Electrical Hazards
and Accidents
Their Cause and Prevention

1

Electrical Systems, Terminology and Components—Relationship to Electrical and Lightning Accidents and Fires

Theodore Bernstein

Investigators of accidents or fires involving electricity and lightning should have a basic knowledge of electricity, its effects as related to such incidents, and the normal and abnormal functioning of electrical components and systems. Electrical power used for homes or industry is generated and delivered to the consumer by power companies whose practices are governed by codes and standards specifically developed to regulate generation and transmission, safety, and performance. The electrical consumer then utilizes this power on premises where electrical systems are designed for local low voltage distribution and utilization. These systems also have to meet specific codes and standards concerned with safety and performance.

CODES AND STANDARDS

The generally accepted code for consumer electrical systems is the National Electrical Code (National Electrical Code 1990), often simply called the Electrical Code or the Code. The National Electrical Code (NEC) is an American National Standard produced and published by a private, nonprofit organization,

1

the National Fire Protection Association (NFPA). This code has no force of law unless it is adopted as law by individual jurisdictions. Because officials in many jurisdictions either do adopt the National Electrical Code or write their own code based on the Code, it is a widely used consensus standard. The National Electrical Code applies to consumer installations from the electrical service entrance where the power enters private property; it does not cover installations exclusively under the control of utilities. Every three years the Code is revised by a group of committees selected from a cross section of industry. The enactment date designated for a newly revised Code is one year after its approval by the National Fire Protection Association to allow time for various jurisdictions to adopt the Code.

The transmission and the distribution of power under the exclusive control of utilities are covered by an American National Standard that also is a consensus standard without the force of law in itself. This standard, called the National Electrical Safety Code (*National Electrical Safety Code* 1990), is revised by industry committees that evaluate and rule on proposed changes; it is produced and published by the Institute of Electrical and Electronics Engineers. Also revised every three years, the National Electrical Safety Code is either adopted or used by all states in developing their own code requirements for utilities.

There are few official governmental standards for nongovernmental privately procured electrical equipment. The most widely used standards for general-use electrical products are those developed by Underwriters Laboratories Inc. (UL), a private, nonprofit organization (Underwriters Laboratories 1990). When the National Electrical Code specifies that "Listed" equipment be used, this usually means equipment that has been manufactured and tested to certain recognized standards, such as those provided by Underwriters Laboratories or other recognized testing laboratories. Other organizations that publish standards for specialized electrical equipment include the National Electrical Manufacturers Association (NEMA) and the Institute of Electrical and Electronics Engineers (IEEE).

ELECTRICAL TERMINOLOGY

One must understand various electrical terms in order to follow any discussion of electrical or lightning problems. A number of these terms are discussed below, and additional basic concepts are presented in Chapter 9.

Voltage

Voltage is a measure of the electrical potential difference *between any two points,* expressed in units of volts (V). It is important to note that the voltage at any specific point always must be measured with respect to that at some other point; one cannot express the voltage at a point without expressing or implying a second point from which the voltage difference is measured. The voltage difference between two points is a factor used to determine the electrical current that will be present in a given electrical path between these points—the higher the voltage is, the greater the current. A good conductor is essentially an equipotential volume with a negligible voltage difference between any two points in the volume or on the surface. Even if the good conductor has a high voltage with respect to another point, it will have an essentially zero voltage difference along its length or within its volume. This explains why squirrels or birds can be standing on high voltage lines without being shocked—the voltage difference between points of contact on the same line is very small, essentially zero, although the voltage between two power lines or from a power line to ground can be quite large.

The magnitude of the voltage difference determines whether the electrical current will break down the insulation between two points. The higher the voltage is, the greater the insulation requirement between conductors. Thus voltage is a major factor in determining insulation requirements.

A voltage difference can exist between two points without producing a current flow if there is no electrical path for current flow between the two points.

Current

Current flow in a conductor, measured in amperes (A), is related to the rate of flow of electrical charge (coulombs per second) in the conductor. The mechanical collisions between charge carriers and the molecular structure of a conductor produce heat, and this heat production is important in determining the size of the conductor required for the safe conduction of a given current. The circuit voltage between conductors and insulation temperature determine the insulation requirements for the conductors. The safe continuous current rating, or the ampacity, of a conductor depends upon the temperature rating for the insulation type, the ambient temperature, and the temperature rise produced by the current flow. The heat produced in a conductor is related to the square of the current in the conductor and the resistance of the conductor.

Resistance and Impedance—Ohm's Law

The resistance (R) or impedance (Z) to current flow in a circuit is measured in ohms. The greek symbol Ω usually is used as an abbreviation for the word ohms. For direct current (DC) circuits or alternating current (AC) circuits with pure resistance loads, dividing the voltage difference across the circuit by the circuit resistance yields the current flow in the circuit. This relationship is called Ohm's law. For alternating current circuits with loads including inductance or capacitance, the total impedance of the circuit is used in Ohm's law to determine the current flow.

Frequency

The current or voltage supplied by power companies usually alternates sinuso-idally with respect to time, alternating between positive and negative peak am-plitudes in a regular fashion at a rate of 60 cycles per second. The unit for cycles per second is called the hertz, abbreviated Hz. The usual frequency supplied by power companies is 60 Hz in the United States and 50 Hz in Europe.

Power and Energy

The power in a circuit, which is measured in watts (W), is related to the product of the effective current in the circuit and the effective voltage across the circuit. For steady state direct current, the effective value for the current or voltage is the constant, steady state value. The effective value for alternating current or voltage is the root mean square (rms) value, or the peak value divided by the square root of two for the special case of sinusoidal alternating current or voltage. Direct current meters usually display the average current, whereas sinusoidal alternating current meters usually are calibrated to indicate rms values. The heating of a device is related to the power consumed by the device. Power may be regarded as a rate of energy use or dissipation; hence, the higher the power, the greater is the rate of energy use or dissipation.

Energy in mechanical and electrical systems is measured in several different units, the most popular ones being the joule (J), the watt-hour (Wh), the British thermal unit (BTU), and the calorie (Cal). One watt is equivalent to the energy consumption of one joule per second. The unit for power often used when motors or engines are considered is horsepower. One horsepower is equal to 746 watts. For a resistance load, power exactly equals the product of effective current and voltage. Power is given in watts when the current is expressed in amperes and the voltage in volts.

For any given fixed value of electrical power to be transmitted across country, a higher voltage will require a lower current. This explains why transmission lines operate at relatively high voltages. The current required is reduced at the higher voltage, and smaller-diameter conductors can be used for long-distance transmission. The higher voltage, however, requires greater insulation. At the voltages used for common transmission lines, this is not a major problem; the conductors are bare, and the necessary insulation is provided by air or by in-sulators at the towers or power poles.

Heating Effects of Electricity

It is easiest to understand the relationship between power dissipation and heating by an example. Suppose the power dissipated, representing the rate of heat energy input, is 100 W. If this power were dissipated in a volume smaller than that of a 100 W lamp, the smaller volume would tend to be higher in temperature than if the power were dissipated in the volume occupied by the 100 W lamp. If 100 W were dissipated in a volume larger than that for the 100 W lamp, the volume would tend to be cooler than that of the 100 W lamp. For a given volume dissipating 100 W, the temperature depends on how well the heat is transferred away by conduction, by convection through air movement, or by radiation to a cooler body. Thus the heating effect or temperature rise is a function of the power dissipation, the volume in which the dissipation takes place, and the cooling action within or around the volume.

Fault

The term fault is applied to a partial or total failure in the insulation or continuity of a conductor. It also is used for the physical condition that causes a device, component, or element to fail. A short circuit is a fault with an abnormal, unintended current path between two points of different potentials, such as a contact or conducting path between the two conductors feeding a lamp. A short circuit between a conductor and ground or a grounded conductor is called a ground fault. An unintended connection between an energized conductor and a grounded metallic chassis or water pipe is an example of a ground fault.

Wire Size, Circular Mils, and Ampacity

The designation for small wire sizes used in the United States is the American Wire Gauge (AWG). Larger sizes are designated directly in units of circular mils (cmil) of cross-sectional area. The AWG relates the gauge number to the cross-sectional area of the conductor in circular mils (see following discussion).

Any other unit of area, such as square inches or square meters, can be readily converted to circular mils by using suitable conversion factors. The cross-sectional area, A, for a conductor with a circular cross section and a diameter d is proportional to the diameter squared:

$$A = Kd^2$$

If the proportionality constant, K, is set equal to $\pi/4$, then:

$$A = \pi d^2/4$$

and the area has the units of the square of the units of the diameter. If the diameter is in inches, then the area is in square inches; if the diameter is in bernsteins, the area is in square bernsteins; and so on.

A different unit of area is obtained by setting $K = 1$, and expressing the diameter of the round wire in mils, where 1 mil equals 0.001 inch. This unit of area is called the circular mil. So if $A = d^2$ cmil, then the area, A, is in circular mils when the diameter, d, is in mils where 1 mil $= 1/1000$ inch. Thus, a wire with a 0.1 inch diameter has a cross-sectional area of 10,000 cmil or 0.00785 in^2 so that one circular mil equals 7.85×10^{-3} in^2. Relationships such as this provide the factors needed to convert from one set of units for area to another, and make possible the use of the circular mil unit for a cross-sectional area of any shape.

The relationship between AWG wire size and cross-sectional area is given in Table 8, Chapter 9 of the Code (*National Electrical Code* 1990). This table also provides data on the DC resistance for different wire sizes. The larger the number of the gauge, the smaller the wire diameter. Wire of size AWG #10 has a cross-sectional area of 10,380 cmil and a diameter of approximately 0.1 inch or 100 mils (actually 101.9 mils). The cross-sectional area of a wire changes by a factor of approximately two for a change of three wire sizes. Thus, AWG #13 has a cross-sectional area of 5,180 cmil, about half that of the 10,380 cmil for AWG #10, whereas AWG #7 has a cross-sectional area of 20,820 cmil, about twice that of AWG #10. Copper AWG #10 has a DC resistance per 1,000 feet of approximately 1 ohm (actually 1.02 ohms). Because this resistance is inversely proportional to the cross-sectional area, AWG #7 has a resistance of approximately 0.5 ohm per 1,000 feet (actually 0.51 ohm), and AWG #13 has a resistance of approximately 2 ohms (actually 2.05 ohms) per 1,000 feet.

The insulation required for a conductor usually is determined by the voltage rating for the conductor and the temperature that the insulation must endure. Higher voltage requires either thicker or better-quality insulation. The current rating or ampacity for a conductor, such as that shown in Table 310-16 of the Code, is determined by the ability of the insulation to withstand the combination of the ambient temperature and the temperature rise produced by the power dissipation in the wire. An insulation often used for general-purpose wiring is polyvinyl chloride (PVC). This insulation has a maximum operating temperature of 75°C (167°F) and a melting temperature of 115°C (239°F).

Additional information concerning ampacity, wire sizing, and conductor insulation requirements is presented in Chapter 3.

GROUNDING

Most electrical systems are grounded; that is, one of the normal current-carrying conductors is connected to the earth. Exposed non-current-carrying metal parts of equipment also are often connected to the grounded conductor and the earth.

There are many misconceptions about electrical grounding. Fault currents to a grounded conductor or object in a home or a plant usually will be carried in grounded metallic conductors with little actual current flow occurring into or within the earth. This is true because the return current path back to the power source for a ground fault to a grounded conductor or object primarily is through metallic conductors. There are currents in the earth, however, when power lines or other energized conductors come into contact with the earth, as occurs when a power conductor is lying on the earth or is in contact with a tree, a crane, a TV antenna, or a person on the ground. Such contacts provide a current path to the earth in place of a path through a grounded conductor. The fact that a conductor contacts the earth does not mean the resistance into the earth is low, as the resistance depends on the type of earth contact and on the earth resistivity at or near the point of contact.

Purposes of Grounding

The reasons for grounding electrical systems are stated in the National Electrical Code (1990), Article 250-1, fine-print note:

> Systems and circuit conductors are grounded to limit voltages due to lightning line surges, or unintentional contact with higher voltage lines, and to stabilize the voltage to ground during normal operations. Systems and circuit conductors are solidly grounded to facilitate overcurrent device operation in case of ground faults. Conductive materials enclosing electrical conductors or equipment, or forming part of such equipment are grounded to limit the voltage to ground on these materials and to facilitate overcurrent device operation in case of ground faults.

Grounding limits undesired voltage excursions by providing a relatively low-resistance pathway into the earth. For example, current surges produced by lightning can have high peak values, the median peak value being 30,000 A. A current transient of this magnitude would produce a high voltage and then potential arcing to any nearby grounded object. The voltage excursion can be limited if a low-resistance pathway is provided to the earth.

Similarly, if a high voltage line of a grounded distribution system came into contact with an ungrounded lower-voltage system, the ungrounded system would rise to the higher voltage. However, if the lower-voltage system were grounded, a large fault current would develop and operate overcurrent devices and clear the fault.

The stray capacitance of AC power lines provides current paths between the lines. Thus, the voltage relative to ground of an ungrounded system can vary widely, depending upon the circuit configuration. A grounding system provides a solid reference for the system voltages relative to the ground.

If a distribution system is not solidly grounded, a fault to ground by an energized conductor may not be detected until a second ground fault occurs. This is so because after the initial fault, the fault current is too small to actuate the overcurrent devices. In a solidly grounded system with low-resistance pathways, an initial fault will produce a high current, and the circuit protection can readily detect and clear the fault.

The practice of grounding exposed metal parts of equipment is called equipment grounding. The inadvertent contact between an energized conductor and an ungrounded exposed metal part will go undetected. The voltage of the exposed metal part will rise, relative to ground, up to the voltage of the energizing conductor. The exposed energized metal surface represents an accident waiting to happen.

In the same fault situation where an energized conductor contacted exposed metal parts, equipment with proper equipment grounding would produce a large fault current. Thus, the circuit protection device would detect and clear the fault. During the time required to clear the fault, the voltage to ground on the exposed metal parts would be limited to the product of the fault current and the resistance to ground of the exposed metal parts.

It is difficult to keep a system ungrounded for a long period of time. Because sneak grounds cannot be avoided, it is best to deliberately ground the system so that at least the location of the ground is known.

Grounding to Earth

Grounding to earth is not a simple undertaking (Biddle Instruments 1981). The earth is composed of poor conducting materials as compared to metallic conductors with fairly low resistivities. (The resistivity is an intrinsic property that determines the resistance for a material sample of a given size and shape; the resistance for any sample of material is directly proportional to its resistivity.) The resistivity of the soil around a ground electrode varies, depending on the nature of the medium, from that of salt water to that of rock, a range of about 1 to 10,000 Ω-m—the resistivity of salt water being about ten million times greater than that of copper. Yet, in spite of relatively high ground resistivities, the resistance to a current exhibited by the earth is substantially zero once the current is in the earth because of the very large cross-sectional area of the earth. The resistance of a conductor is inversely proportional to the cross-sectional area of the conductor—the larger the cross-sectional area is, the smaller the resistance.

The resistance of a ground connection to the earth is primarily the resistance of the earth in the immediate vicinity of the grounding electrode where the current leaves the grounding electrode to enter the earth. Article 250-81 of the

National Electrical Code lists preferred grounding electrodes with large areas of contact with the soil—in other words, those with usually low grounding resistance:

- A metal underground water pipe in direct contact with the earth for 10 feet or more.
- The effectively grounded metal frame of a building.
- An electrode encased by at least 2 inches of concrete located near the bottom of a concrete foundation or footing in direct contact with the earth.
- A ground ring encircling a building or structure.

If none of the above systems for electrodes is available, then the electrodes in Article 250-83 may be used:

- Other local metal underground systems or structures.
- Rod and pipe electrodes at least 8 feet in length.
- Plate electrodes that expose at least 2 square feet of surface to exterior soil.

It is important to realize that a driven ground rod or a plate electrode usually is the poorest of the various types of grounding electrodes and should be used alone only as a last resort.

Driven Ground Rods

There is a serious misconception that driven ground rods provide a low resistance to ground. The resistance exhibited by a driven ground rod electrode has three components:

- Metallic resistance of the electrode itself.
- Contact resistance at the surface between the metallic electrode and the earth.
- Resistance of the earth immediately around the electrode.

Of these three components, usually the last, the resistance of the immediately surrounding earth, is by far the largest. The resistance of the metallic electrode and the contact resistance are much smaller than the earth resistance contribution.

It is not unusual to drive an 8-foot ground rod in some locations with high soil resistivity and find a resistance to ground of 2,000 ohms or more. In some low-resistance soils, a resistance to ground of 5 ohms or less can occur. The only National Electrical Code (1990) reference to ground resistance for a made electrode is found in Article 250-84, where it is stated that if the ground resistance for a made electrode is not 25 ohms or less, the made electrode must be augmented by another grounding electrode of any type permitted in the Code. It should be noted that no ground resistance for the new required parallel combination of electrodes is specified in the Code. Therefore, ground resistance for the parallel combination need not be measured—but it may be greater than 25 ohms.

A longer ground rod will decrease the resistance to earth by about 40 percent when the ground rod length is doubled. The ground rod diameter has little effect on the resistance to ground, so increasing the rod diameter does not decrease the resistance to ground significantly.

Care must be exercised in paralleling ground rods to decrease resistance. If the resistance of each of two ground rods to earth measured individually is R ohms, paralleling the two will not halve the ground resistance. The closer the rods are together, the closer the resistance of the two rods in parallel will be to that for one of the rods. To illustrate this fact, consider the following example. Visualize a thick ground rod of a certain length and diameter driven into the ground. It will have a certain resistance to ground. Then slit the rod in half lengthwise to make two ground rods. The resistance to ground for each half will be about the same as for the original rod, as only the approximate diameter of the rods has changed. If the two halves now are joined at the earth's surface, the two rods are in parallel, as this is the original configuration with just a lengthwise slit in the original rod. The resistance to ground will be about the same as for the original single rod, even though it is now equivalent to two rods in parallel. When the rods are far apart, at least one rod length, then the rods tend to act as parallel paths to ground. Well-separated rods will reduce earth resistance to about 60 percent of a single rod for two rods, 40 percent for three rods, and 33 percent for four rods. Chemical treatment of the soil around the rod, as long as the treatment lasts, also can be used to reduce earth resistance because most of the resistance is caused by the resistivity of the soil near the rod.

Measurement of Earth Electrode Resistance

Measurement of the grounding electrode resistance usually requires special equipment. It is nonsense to try to measure the resistance to ground by using an ohmmeter with one lead attached to the electrode and the other driven into the ground. The low-level voltages and currents, as well as the high resistance to ground of the lead in the ground, give meaningless results.

The usual method for measuring the earth resistance for a ground rod is called the three-terminal method. A small current-collecting probe is driven into the ground at a distance of about 100 to 120 feet from the driven ground rod under test. (See Figure 1.1.) A known current is injected at the ground rod and collected by the current probe.

A voltage probe is inserted at any point along the surface of the earth to measure the voltage between the probe and the ground rod. A rapid change in voltage with respect to distance is observed at points near either the ground rod

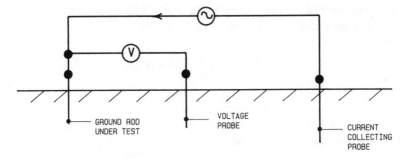

Figure 1.1. Method of measurement of the earth's resistance.

or the current-collecting probe. Somewhere in the center region between these two electrodes the variation of voltage with distance becomes almost negligible. This region of small voltage change indicates that the current is essentially diffused into the earth. By using Ohm's law and dividing the voltage measured between this region and the ground rod under test by the injected current, the resistance of the ground rod under test is determined.

Calculation and Variation of Ground Resistance around a Grounding Electrode

To obtain an order of magnitude for the ground resistance of a grounding electrode such as a plate on the ground or a ground rod, it is assumed that the ground contact can be approximated by a conducting hemisphere of radius r_0 meters buried in the earth, as illustrated in Figure 1.2.

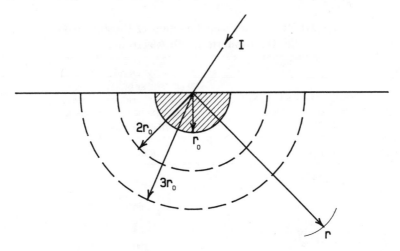

Figure 1.2. Hemispheric conducting electrode buried in the earth.

The resistance of the earth between the conducting hemisphere of radius r_0 and another conducting hemisphere of radius r is:

$$R = \frac{\rho}{2\pi}\left[\frac{1}{r_0} - \frac{1}{r}\right]\ \Omega \tag{1-1}$$

where ρ is the resistivity of the earth in ohm-meters, and r_0 and r are expressed in meters (Tagg 1964, 91). As $r \to \infty$, the resistance between the hemisphere of radius r_0 and the entire earth is:

$$R_\infty = \frac{\rho}{2\pi r_0}\ \Omega \tag{1-2}$$

Most of the resistance occurs in the region near the hemispherical electrode. This can be shown by inserting the resistance between the hemispherical electrode and the entire earth from equation (1-2) into equation (1-1), so that:

$$R = R_\infty\left[1 - \frac{r_0}{r}\right]\ \Omega \tag{1-3}$$

and:

$$\frac{R}{R_\infty} = \left[1 - \frac{r_0}{r}\right]\ \Omega \tag{1-4}$$

Table 1.1. Resistance as a Function of Distance from the Hemispheric Earth Electrode.

RATIO OF RADIUS TO POINT IN EARTH TO RADIUS OF EARTH ELECTRODE	RESISTANCE BETWEEN EARTH ELECTRODE AND HEMISPHERE OF RADIUS r AS A PERCENTAGE OF RESISTANCE BETWEEN EARTH ELECTRODE AND HEMISPHERE WITH INFINITE RADIUS
(r/r_0)	(R/R_∞) (%)
1	0
2	50
3	66.7
5	80
10	90
∞	100

Table 1.2. Grounding Electrode Resistance for a Hemisphere with a 7.5 cm Radius (Approximately 3-inch Radius).

MATERIAL	RESISTIVITY (Ω-m)	R (Ω)
Salt water	1	2.12
Ashes, cinders	24	50.9
Clay, shale	40	84.8
Some soils	100	212
Gravel, sand, granite	1,000	2,120

Using equation (1-4), the percentage of the resistance between the electrode and the entire earth that occurs between the electrode radius r_0 and a given distance r can be calculated. In Table 1.1, 50 percent of the earth resistance for such a grounding electrode occurs within one radius of the electrode surface.

Using equation (1-2), an order or magnitude for the resistance of an earth electrode can be obtained. Some calculated resistance values for different types of material about the electrodes are shown in Table 1.2.

Additional discussions of grounding and resistivity are presented in Chapters 4 and 9.

THE 60 HZ GENERATION, TRANSMISSION, AND DISTRIBUTION SYSTEM

The usual electrical system for general use in homes and light industry is the single-phase, alternating, sinusoidal, 120/240 V, 60 Hz electrical system shown in Figure 1.3. Single-phase, alternating, sinusoidal voltage or current indicates a voltage or current that is varying in time between positive and negative values in a sinusoidal fashion. When the voltages between any two supply points have a time variation such that the voltages always reach their peak amplitude at the same time, the supply is called single-phase. The term 120/240 V identifies the two nominal voltages available. The 120/240 V voltages are the effective or root mean square (rms) voltages between conductors. The peak or maximum amplitude voltagesbetween conductors would be found by taking the square root of two times the effective voltage. The effective or root mean square voltage is the usual value read on an AC voltmeter and is used for circuit calculation. Magnitudes for sinusoidal, alternating voltages or currents usually are expressed as the effective or rms value and not the peak value. For industrial or commercial establishments, voltages of 480 V rms or higher may be supplied.

Power often is generated by utilities at voltages of 10 to 20 kV as three-

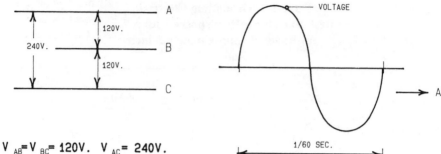

$$V_{AB} = V_{BC} = 120V. \quad V_{AC} = 240V.$$

(a) SINGLE PHASE VOLTAGE   (b) VOLTAGE WAVEFORM

Figure 1.3. Single-phase, 120/240 V, 60 Hz system. (a) Single-phase voltage. (b) Voltage waveform.

phase, 60 Hz power. The term three-phase indicates that there are three conductors associated with the system, with the nominal voltage indicated being the rms voltage between any two of the three conductors. Thus, a 13.2 kV, three-phase system has an effective alternating voltage of 13.2 kV between any two of the three conductors. The system is called three-phase because there is a phase

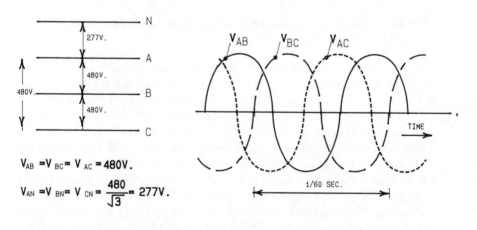

$$V_{AB} = V_{BC} = V_{AC} = 480V.$$

$$V_{AN} = V_{BN} = V_{CN} = \frac{480}{\sqrt{3}} = 277V.$$

(a) THREE-PHASE LINE-TO-   (b) THREE-PHASE SINUSOIDAL
     LINE VOLTAGE RELATIONSHIP   VOLTAGE WAVEFORMS

Figure 1.4. Three-phase, 277/480 V, 60 Hz system. (a) Three-phase line-to-line voltage relationship. (b) Three-phase sinusoidal voltage waveforms.

or time difference between the instants when each of the three alternating voltages reaches a peak value. Thus the peak voltages between the conductors do not occur at the same instant of time. This is illustrated in Figure 1.4 for one cycle of the 60 Hz voltage.

A voltmeter displaying the voltage between any two of the three phases would not indicate the time or phase difference between the phases. The three-phase conductors usually are not grounded, although there are some systems where one of the phase conductors is grounded. Many systems use a grounded Y connection, as shown in Figure 1.5, with a fourth wire, the neutral, grounded

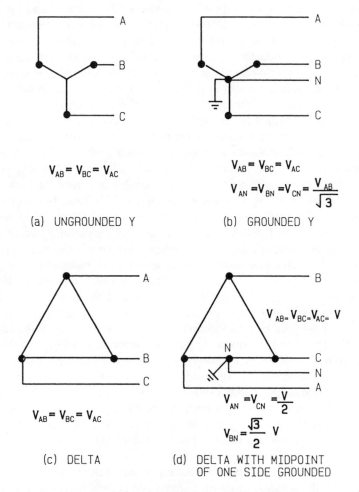

$$V_{AB} = V_{BC} = V_{AC}$$

(a) UNGROUNDED Y

$$V_{AB} = V_{BC} = V_{AC}$$

$$V_{AN} = V_{BN} = V_{CN} = \frac{V_{AB}}{\sqrt{3}}$$

(b) GROUNDED Y

$$V_{AB} = V_{BC} = V_{AC}$$

(c) DELTA

$$V_{AB} = V_{BC} = V_{AC} = V$$

$$V_{AN} = V_{CN} = \frac{V}{2}$$

$$V_{BN} = \frac{\sqrt{3}}{2} V$$

(d) DELTA WITH MIDPOINT
OF ONE SIDE GROUNDED

Figure 1.5. Typical three-phase configurations and voltage relationships. (a) Ungrounded Y. (b) Grounded Y. (c) Delta. (d) Delta with midpoint of one side grounded.

so that the voltage between any phase conductor and ground is the phase-to-phase voltage divided by the square root of three. A 13.2 kV, three-phase system has a voltage of 7.62 kV to ground. Three-phase is used to provide more efficient transmission of larger powers. In addition, large motors run more smoothly and with greater efficiency when they are operated from three-phase. Two delta configurations and an ungrounded Y configuration for three-phase systems are shown in Figure 1.5. The simple delta configuration may have one of the line conductors grounded. The delta with the midpoint of one side grounded often is used in three-phase, 240 V line-to-line voltage systems, as this configuration also provides a 120/240 V single-phase power for loads. Information concerning the type of configuration usually is only important in accident cases for determining the voltage between any two conductors or between a conductor and ground. In most accidents or fires, the initial contact or fault is a single-phase fault between two conductors or between a conductor and an electrical ground; therefore, only the single-phase voltage between the contact points is significant.

Utilities may generate power at voltages of 10 to 20 kV. The upper limit on voltage is determined by properties of the insulation in the generator. Lower generated voltages require larger conductors in the generator. These voltages are raised by transformers to standard three-phase, line-to-line voltages, such as 69, 115, 138, 230, 345, or 500 kV for transmission over great distances. The higher voltage and the lower current permit the use of smaller conductors. The voltage then is stepped down by using transformers, possibly in several stages, to 4.16, 12.47, 13.2, 24.94, or 34.5 kV line-to-line or 2.4, 7.2, 7.62, 14.4, or 19.9 kV line-to-ground for distribution to the ultimate user (ANSI C84.1-1989). Knowledge of these voltages facilitates the determination of the potential current which can occur in the event of contact with the lines. The conductors that supply the transformers that reduce the distribution voltage for final consumer use are called the primary feeders. Common primary distribution voltages to ground in rural or residential areas are 7.2 or 2.4 kV, the latter used for older systems. Some newer systems use 19.9 kV to ground voltages for the primary feeders. The lower voltage supplied by the secondary of the transformer, such as the 120/240 or 480 V supplied to a consumer, is called the secondary voltage.

The primary conductors frequently are involved in accident cases. These conductors usually are bare—sometimes referred to as air-insulated—and obtain their insulation at the power poles with insulators. Usually it is the primary feeder that is involved in cases where a crane contacts a conductor, or someone with a television antenna or an aluminum ladder contacts the conductor. The primary conductors near homes and industry rarely are insulated. In some installations, the primary and the secondary conductors are buried.

Transmission and distribution systems usually are under the control of a utility. Governed by the National Electrical Safety Code (1990), these systems usually are three-phase and use higher voltages than those used in homes or industry.

The value of the voltage that supplies a ground fault in a three-phase system usually is the line-to-line voltage divided by the square root of three. Thus, a 69 kV, three-phase system has a voltage to ground of 39.8 kV. This is the value of the voltage between the phase conductor and the ground, and it is an important quantity because it determines the fault current that will flow with a given ground fault resistance. Most electrical accidents involving power lines involve contact between one of the line conductors and the ground.

THE 60 HZ CONSUMER ELECTRICAL SYSTEM

The configuration of the electrical system serving a home, office, industrial plant, or retail establishment is determined by the electrical load requirement. Most homes are supplied with a 120/240 V single-phase supply, as are many smaller office and retail establishments. Larger commercial buildings and industrial plants are supplied with a higher-voltage three-phase system that can supply the larger electrical load demands as well as provide for the usual 120 V equipment.

The 120–240 V (Primarily Residential) Electrical System

A typical service entrance for a building is shown in Figure 1.6, where the power company's primary single-phase voltage, 7200 V in this case, is converted to the 120/240 V secondary voltage for use in the building. The transformer often is outside the building, and only the three secondary conductors are brought into the building, either overhead by a service drop or underground by a service lateral. One of these three conductors, the grounded conductor, is grounded to earth at the transformer and at the service equipment grounding electrode. The service equipment, which is located near the building service entry, consists of the necessary circuit breaker or switch and fuses with the accessories necessary to disconnect power to the building. Service equipment grounding is accomplished by connection to a metallic water main system, other forms of extensive buried metal with good contact with the earth, or a ground rod. The voltage between the two ungrounded lines is 240 V, and the voltage from either ungrounded conductor to the grounded conductor is 120 V.

The grounded conductor, sometimes referred to as the neutral conductor, always has white-colored insulation when routed inside a building or on the load side of the service equipment. The grounded conductor in a building is insulated and is grounded only at the service equipment, except for a few permitted exceptions such as the grounding of clothes dryers or electric ranges. Ungrounded conductors cannot have an insulation color of white or green, as these colors are reserved for grounded or grounding conductors. The nominal 120 V sometimes is referred to as 110 V or 115 V.

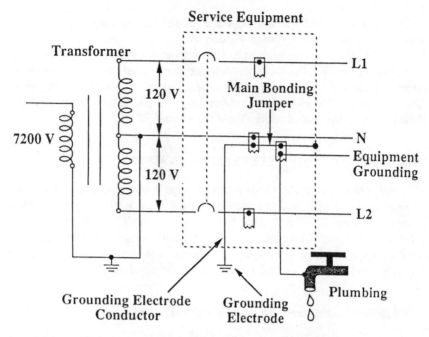

Figure 1.6. Typical 120/240 V single-phase service entrance.

The equipment grounding conductor shown in Figure 1.6 is bonded to the grounded conductor and the service equipment cabinet at the service entrance. This conductor, either bare or with a green-colored insulation, is used for grounding all exposed metal parts that might become energized in the event of an electrical fault in the equipment. Unlike the grounded conductor, the equipment grounding conductor can be grounded at many places as long as there is an effective metallic path for fault current back to the neutral at the service equipment. Without an equipment grounding conductor, an ungrounded 120 V conductor might contact a metal enclosure and not operate any circuit protection because the enclosure would not be grounded; so there would be no current in the circuit. The enclosure would be at 120 V with respect to ground. A person touching the enclosure and a grounded object, such as a grounded electrical device or plumbing, could provide a path for current through his or her body from the faulted equipment to ground.

Typical interior wiring connections in a residential or commercial building are shown in Figure 1.7. The double pole circuit breaker shown is the main disconnect for the system. A 150 A service would have a 150 A main circuit breaker. The ratings of the remaining circuit breakers are selected to match the ampacity for the smallest-size branch circuit conductor that each supplies. A branch circuit consists of the circuit conductors between the final overcurrent device protecting the circuit and the outlet or outlets. An outlet is a point on the wiring system, such as a receptacle, light fixture, or switch, where current is taken to supply utilization equipment. An AWG #14 wire protected with a 15 A overcurrent device is the smallest-size conductor permitted for branch circuit wiring. The lamp shown is supplied with 120 V; one terminal is connected to the white insulated grounded conductor, while the other terminal receives power through the switch from the 15 A circuit breaker. The wire insulation for this ungrounded conductor has any color other than green or white. The 120 V heater

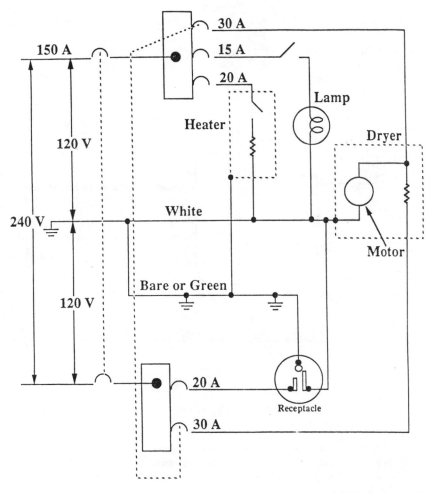

Figure 1.7. Connections for various types of residential/commerical 120/240 V single-phase loads.

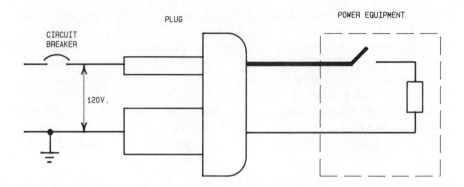

(a) PROPER POLARITY WITH POLARIZED PLUG

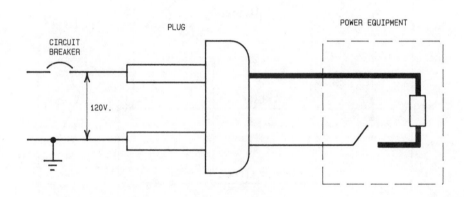

(b) REVERSED POLARITY AND NON-POLARIZED PLUGS

NOTE: HEAVY LINES SHOW CIRCUIT ENERGIZED AT 120V.
       TO GROUND WHEN THE SWITCH IS OFF.

Figure 1.8. Two-prong polarized plug and nonpolarized plugs. (a) Proper polarity with polarized plug. (b) Reversed polarity and nonpolarized plugs. *Note*: Heavy lines show circuit energized at 120 V to ground when the switch is off.

is supplied by a 20 A circuit breaker, and the metallic heater case is grounded by the bare or green-colored equipment grounding conductor. The clothes dryer heating element, being a large electrical load, is operated at 240 V to reduce current requirements, and a 120 V motor turns the dryer drum. The dryer chassis, by special exception in Article 250-60 of the National Electrical Code (1990), uses the grounded conductor for equipment grounding.

The grounding-type receptacle shown should be wired for proper polarity with the ungrounded conductor connected to the shorter socket with the brass or copper-colored screw connection. The grounded conductor is connected to the longer socket with the white or silver-colored screw connection. The circular socket for the attachment plug's third prong is connected to the bare or green insulated equipment grounding conductor or to a conduit system that provides an effective equipment grounding path back to the service entrance. When a three-prong grounding-type plug is inserted into the receptacle, the equipment supplied by the power cord will have 120 V for operation through the two parallel blades, and equipment grounding will be provided by the equipment grounding pin.

A two-bladed polarized plug, sometimes used for equipment that does not require equipment grounding, can be plugged into the receptacle only one way because one of the blades is wider than the other. The power switch in the device being powered always will open the ungrounded conductor to leave the circuit on the load side of the switch at ground potential when the switch is open and the device is off. Figure 1.8 shows that with a properly polarized plug most of the circuit on the load side of the equipment switch will be at ground potential when the switch is off; with reversed polarity, the circuitry on the load side of the equipment switch is at line potential when the equipment switch is off. The equipment will operate satisfactorily with proper or reversed polarity; proper polarity only provides an added safety features in the event that circuitry in the equipment is contacted when the equipment is turned off. The three-prong grounding-type plug is inherently polarized and can be inserted in the socket only one way.

Higher-Voltage Consumer Electrical Systems

For consumer electrical loads requiring more power than can reasonably be supplied by the single-phase 120/240 V electrical system, or where three-phase power is necessary for loads such as motors, other systems often are provided. Commonly used systems are 208Y/120 V and 480Y/277 V systems with a line-to-line voltage of 208 or 480 V and a line-to-neutral or ground of 120 or 277 V. Examples of these systems are illustrated in Figure 1.5. Another frequently used system has the secondary circuit connected in delta with the midpoint of one side grounded. With a three-phase 240 V line-to-line system in this configuration, 120 V is available for single-phase loads from two of these lines to ground. (See Figure 1.5d.) The third line, sometimes called the "wild leg," has a voltage of 208 V to ground, and it is not permissible to connect a 208 V load from this line to ground. Large plants may have systems operating with voltages in the kilovolt range.

ELECTRICAL COMPONENTS

Fuses and Circuit Breakers

The ungrounded conductor is protected by a fuse or a circuit breaker that primarily protects the wire insulation from overheating because of overcurrent. The rating of the fuse or the circuit breaker selected depends on the ampacity of the wire it is protecting. The fuse or the circuit breaker is designed so that it will operate in the event that there is excessive current in the ungrounded conductor. It has a current-versus-time operating characteristic such that the conductor insulation will not be overheated and damaged by the current. A fuse or a circuit breaker has an inverse current time operating relationship: the higher the current, the faster the operation. A typical 15 or 20 A circuit breaker will carry its rated current indefinitely. The breaker will actuate in less than 1 hour at a 125 percent rating and in less than 2 minutes at a 200 percent rating (UL 489 1986). At a 300 percent breaker rating, the usual trip time is between 5 to 35 seconds (Westinghouse 1976).

Fuses have similar characteristics. Fuses or circuit breakers offer no primary protection from electrical shock, nor do they provide protection from fires caused by high resistance or arcing faults. Lethal currents are on the order of 0.05 A, well below the usual fuse or circuit breaker operating points. Fires can be started at currents below 15 A by the heat of an arc, the overheating of a small conductor, an overheated component, or the heat at a loose terminal or connection. A fuse or a circuit breaker primarily will protect the wiring in the wall, nothing more.

Plugs and Receptacles

Wall outlet receptacles installed after 1962 are of the grounding type with sockets for the two parallel blades and a grounding pin on the attachment plug. The three-prong grounding-type attachment plug used to energize power cords for equipment provides 120 V power between the two parallel blades when inserted into a receptacle. The equipment grounding pin on the plug is connected, by a green wire in the power cord, to all exposed metal parts of the equipment. Under normal operating conditions, there is no current in this third, equipment grounding pin. If there is an insulation failure in the equipment, and the exposed metal parts make a connection to an energized conductor, the equipment grounding conductor provides a return path for the electricity. Sufficient current will flow in the energized conductor to actuate the fuse or the circuit breaker protecting the ungrounded conductor. The equipment then is deenergized because of the fault current.

In addition, the equipment grounding conductor will tend to keep the chassis at ground potential even if there is not sufficient ground fault current to trip the overcurrent device protecting the circuit. This is possible if the ground fault in the equipment has a high internal resistance in the fault path. The equipment grounding conductor carries no current and plays no part in the normal operation of the equipment, but is an important safety feature. Without the equipment grounding conductor, a fault in the equipment, where the energized conductor comes in contact with exposed metal parts, might or might not causae the circuit breaker or the fuse to operate; its activation would depend upon there being a sufficiently low-resistance path to the ground from the exposed metal parts. Equipment with such a fault resting on a wooden table would not have a sufficiently low-resistance path to ground from exposed metal parts to activate the overcurrent protection device. The equipment could continue to function normally with its exposed metal parts at a voltage of 120 V. Someone simultaneously touching this exposed metal and a grounded object, such as a faucet or a grounded tool, could provide a current pathway and receive a lethal shock.

The equipment grounding conductor and pin arrangement resembles a safety valve for a boiler; it is not needed for normal operation but is an essential safety feature in the event of a failure. Equipment grounding never should be defeated by breaking off the equipment grounding pin or by using an adapter for conversion from a three-prong to a two-prong plug without properly grounding the adapter to the grounded screw in the receptacle face plate. The adapter, which allows a grounding-type plug to be inserted into a two-bladed, ungrounded receptacle, will effectively disconnect the grounding pin of the plug from equipment grounding unless the tab on this adaptor is connected to a proper ground.

Ground Fault Circuit Interrupter

One way to prevent lethal electrical shocks is to use a ground fault circuit interrupter (GFCI). The GFCI operates as a device that interrupts power in the event of an accidental low-current-magnitude circuit to ground, other than a path through the grounded conductor. Most electrocutions occur when an individual contacts an energized, ungrounded conductor or energized defective equipment and a ground other than the grounded conductor. The GFCI detects low-level ground faults and opens the circuit.

As discussed in Chapter 3, the human body and contact resistance tend to limit the current involved in electric shocks. Current on the order of 50 mA can cause an electrocution, and a conventional circuit breaker operating at 15 to 20 A could never provide protection for such low-current shocks. A GFCI can provide protection, as it trips at a current to ground of less than 6 mA (UL 943 1985). As shown in Figure 1.9, the GFCI compares the current in the ungrounded conductor to that in the grounded conductor for conductors routed to an electrical device. This comparison is implemented by routing both the ungrounded conductor and the grounded conductor through a magnetic core. The voltage produced in the sense coil wound on the core will be proportional to the difference in current in the two conductors.

When there is a difference in current of 6 mA, the coil voltage is sufficient to operate the interrupter for that line and to deenergize the circuit within 5.6 seconds. At 264 mA the GFCI will trip within 0.025 second. The basic concept is that the unbalance in current is caused by a ground fault, an unintended path to ground. The current trip times for the GFCI are designed to be in the safe range for human shock. This ground fault could be a person touching the energized conductor and ground. At any level of shock above the 6 mA level, an individual could be frozen to the circuit and not be able to let go of an energized conductor. By tripping at 6 mA, the GFCI ensures that an individual will be able to let go of the circuit and not receive a lethal shock. The shock

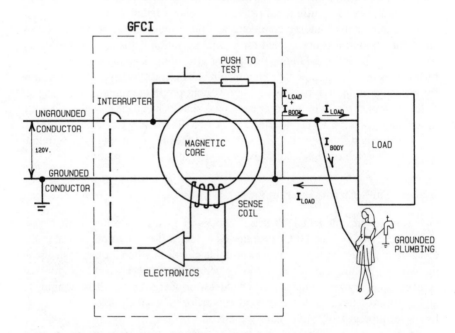

Figure 1.9. Schematic diagram for operation of the ground fault circuit interrupter (GFCI).

might be painful, but it will not be lethal. The resistance of the human body and not the GFCI would tend to limit the ground fault current. For low-resistance ground faults, such as would occur if someone were to short-circuit the GFCI with a screwdriver, the ground fault current can be very high.

The push-to-test button incorporated in the ground fault circuit interrupter creates an unbalance in current through the magnetic core of less than 6 mA. This causes the GFCI to trip and provides a test of the device's operation.

Double Insulation

Double insulation, of electrical tools and appliances, implies that the insulating system between internal energized conductors and any possible point of external contact consists of both functional and protecting insulation, with the two physically separated (UL 1097 1983). Functional insulation is the insulation necessary for the proper operation of an appliance, such as the winding insulation of a motor or a transformer. Protecting insulation is an independent insulation that provides protection against electric shock in case of failure of the functional insulation; an enclosure of insulating material is an example of protecting insulation.

Double-insulated appliances are required only to have a two-bladed plug. Any exposed metal parts have a protecting insulation inside. For example, the exposed metal chuck on a drill has an internal protecting insulator, not part of the motor functional insulation, which insulates the exposed metal chuck from the internal metal parts of the electric motor driving the chuck.

THE ELECTRIC ARC

The properties of the electric arc are important in accident investigation. Arcs can cause or be caused by equipment failure and fires; they also can play a part in power line contact accidents. Depending on its size and available energy, the electric arc can be a source for fire ignition at currents well below 15 A. The temperature of an electric arc at these current levels is on the order of 2000°C to 4000°C (3600°F to 7200°F). (Cobine 1958, 291). Such arcs can cause damage *if sufficient energy is available in them.*

The breakdown strength for air that can initiate an arc depends on the shape of the electrodes and the waveform of the applied voltage. A typical value used is 30 kV per centimeter or 76.2 kV per inch (Cobine 1958, 174). This means that there must be a voltage difference of 30,000 V for each centimeter of gap length, or 76,000 V for each inch of gap length, to initiate an arc through air. A fundamental property of arcs is that no matter how small the air gap is between two conductors, there can be no breakdown across an air gap if the voltage is below approximately 300 V (Cobine 1958, 165). After the arc is initiated, only 20 V per centimeter or 50.8 V per inch is required to maintain it (Golde 1973, 178); thus, once initiated, the arc will be maintained over a much larger air gap length than the original initiation distance. The arc voltage is practically independent of arc current and depends primarily on arc length. As an example, an arc welder with an 80 V open circuit voltage cannot initiate an arc across an air gap; but when the electrodes touch and are separated, an arc is initiated so that the voltage then can maintain the arc. These facts help refute certain assertions, such as that a crane was a foot or two away from a 7200 V line when electricity arced to the crane. The crane actually must have been within 0.1 inch of the line, or more likely touched the line, and the arc occurred when the crane boom pulled away.

It is more difficult to initiate an arc through moist air than through dry air because more voltage is required across a given-length air gap to initiate an arc through moist air than through dry air (Cobine 1958, 182). Note that this applies to an arc through air and not an arc across a surface.

Arcs can be initiated at lower voltages, well below 300 V, across a surface or through some material when there is a low-resistance path permitting currents to heat the material and allow an arc to be initiated over the surface. Insulated conductors in a fire can arc, without touching, at a voltage of 120 V when the insulation chars, permitting currents through the damaged insulation and hot gases that then can lead to an arc. Circuit protection may not operate promptly to limit arc damage. An arc can act as a somewhat high-resistance fault and limit the current because of the voltage drop necessary to maintain the arc. Hot gases from a fire can allow an arc through air at voltages well below 30,000 V per centimeter. Other arcs can be initiated at low voltages if a worker bridges energized conductors with devices such as a screwdriver or a metal tape and there is sufficient electrical energy to maintain the arc.

REFERENCES

ANSI C84.1-1989. *Voltage Ratings for Electric Power Systems and Equipment (60 Hz)*. New York: American National Standards Institute.

Biddle Instruments. 1981. *Getting Down to Earth. A Manual on Earth-Resistance Testing for the Practical Man. Fourth Edition*. Blue Bell, PA: Biddle Instruments.

Cobine, J. D. 1958. *Gaseous Conductors*. New York: Dover.

Golde, R. H. 1973. *Lightning Protection*. New York: Chemical Publishing.

National Electrical Code 1990. NFPA 70 1990 Edition. Quincy, MA: National Fire Protection Association.

National Electrical Safety Code. ANSI C2-1990. New York: Institute of Electrical and Electronics Engineers.

Tagg, G. F. 1964. *Earth Resistance*. New York: Pitman.

UL 489 1986. *Standard for Safety. Molded-Case Circuit Breakers and Circuit Breaker Enclosures*. Northbrook, IL: Underwriters Laboratories Inc.

UL 943 1985. *Standard for Safety. Ground-Fault Circuit Interrupters*. Northbrook, IL: Underwriters Laboratories Inc.

UL 1097 1983. *Standard for Safety. Double Insulation Systems for Use in Electrical Equipment*. Northbrook, IL: Underwriters Laboratories Inc.

Underwriters Laboratories. 1990. *Standards for Safety Catalog*. Northbrook, IL: Underwriters Laboratories Inc.

Westinghouse 1976. *Application Data 29-160*. Beaver, PA: Westinghouse Electric Corp.

2

Physiological Effects of Electricity—Relationship to Electrical and Lightning Death and Injury

Theodore Bernstein

The physiological effects of electrical shock vary from shocks so small that they are not even perceived to severe shocks producing severe tissue damage or even death. The electrical parameters important in evaluating electrical injury cases are the magnitude and the electrical waveform, such as alternating current, direct current, pulse, or lightning discharge. The usual mode of death in electrocution is cardiac arrest or, for high energy shocks, severe tissue damage. The term electrocution refers only to death by electricity and not to electrical shock in general. The physiological response to a 60 Hz, sinusoidal electrical shock will be described initially. Then the physiological response to other waveforms that produce similar effects will be considered. Finally, this information will be used to discuss several case histories of actual electrical accidents.

SHOCKS INVOLVING 60 HZ, SINUSOIDAL WAVEFORM VOLTAGES

Human Body Electrical Impedance

The effects of 60 Hz, sinusoidal electric shocks are related to the effective or root mean square value of the current and the duration of the shock. The voltage is important only insofar as the voltage and the impedance of the circuit together determine the current according to Ohm's law. Thus it is necessary to know the approximate human body impedance so that the shock current can be accurately estimated when the voltage difference is known. An individual touching two points with different voltage levels completes an electrical circuit between the points. The resulting current depends upon the magnitude of the total circuit impedance, which is a series combination of the body skin contact and the body volumetric impedances between the contact points (IEC 479-1 1984).

The volumetric impedance depends on the amount and the types of body tissues in the current path. For alternating voltages it is the impedance rather than just the circuit resistance that determines the magnitude of the current. The minimum possible value of volumetric impedance between two points of contact is significant. This value determines the maximum body current possible for a given voltage difference. The actual current always will be less than this because there always is skin contact impedance present.

The skin contact impedance includes the impedance of the skin and the contact impedance at the skin interface. The voltage, frequency, current duration, contact surface area, contact pressure, skin condition, and moisture level all influence the skin contact impedance. It varies widely from thousands of ohms with dry skin and a small contact area to negligible values when the skin's integrity is compromised with lacerations or by heating effects of the current.

The resistivities within the human body for bone, fluids, and various tissues vary widely (Geddes and Baker, 1967). The body impedance is primarily resistive, with a minimum value of about 500 ohms between any two limb extremities. This value would correspond to an average resistivity for the body similar to that of a semiconductor or salt water, or about 1 ohm-meter. The 500-ohm minimum resistance would indicate that the maximum current possible for a limb-to-limb shock, such as hand-to-hand or hand-to-foot, would be 240 mA for a 120 V shock, or 480 mA for a 240 V shock. It is important to note that because of the skin contact impedance actual currents never will be this high. In addition, if gloves, shoes, or other clothing is in the current path providing additional impedance, the potential current available is reduced further. Using the previous value of resistivity, the lowest body resistance directly across the chest is on the order of 50 to 100 ohms.

Some typical total impedances, limb-to-limb with large contact areas, have been measured, in tests conducted on human volunteers and cadavers (Biegel-meier 1985). The total impedance depends on the applied voltage. For the test group, only 5 percent of the group had impedances below 1,750 ohms at 25 V, 1,125 ohms at 120 V, 750 ohms at 700 V, and 650 ohms above 1,000 V. This finding would tend to indicate that the minimum value of 500 ohms between any two limbs is a conservative value.

Threshold of Perception and Startle Reaction

The threshold of perception for a finger tapping contact at 60 Hz is approximately 0.2 mA. A current of 0.36 mA could be perceived by 50 percent of a group of men, whereas 50 percent of a group of women perceived 0.24 mA (Kahn and Murray 1966). Such low-level shocks are not dangerous, but shocks above the threshold level can present a hazard if the low-level shock startles an individual. The resulting involuntary motion might lead to a harmful accident.

Scientist at Underwriters Laboratories (Smoot and Stevenson 1968) studied a group of men and women to determine the level of current that would produce a startle reaction, defined herein as an uncontrolled muscular reaction resulting from an electrical shock. During the test, each subject sat in a test booth with one hand in a bucket of salt water that served as one electrode. The subjects performed tasks such as filling a metal container with rice, moving the container about, and then pouring out the rice with the free hand. As the free hand and arm were moved about the booth, surprise shocks with different current levels were delivered at various contact points. The reactions to the shock then were evaluated. The results indicated that if the shock current was below 0.5 mA, a shock might be perceived, but no uncontrolled startle reaction resulted. The women seemed to have a greater sensitivity to shock than the men. The threshold level for a startle reaction was established to be 0.5 mA, which is used as the maximum allowable leakage current in testing new appliances (ANSI C101.1 1986) and for ground fault circuit interrupters.

Shocks above but near the startle reaction level can be dangerous. Such low-current shocks in the body will not cause physiological harm, but the startle reaction may be a hazard if it leads to a fall or to contact with dangerous equipment.

Let-Go Current

A current level somewhat higher than the threshold of perception is called the let-go current level (Dalziel and Massoglia 1956). This level is important because with a current of this magnitude flowing in an individual's hand and arm, the hand will involuntarily close and grasp the electrified contact that the palm or fingers were touching. The individual is involuntarily held or frozen to the energized conductor through the grasp and cannot "let go" unless the power is turned off or the individual is physically removed from circuit contact. If contact is not broken, contact resistance may decrease because of perspiration, tearing of the skin, or a tighter grasp, allowing larger and lethal currents to pass through the body. Shocks at the let-go current level are quite painful although not usually lethal. The let-go current for women is lower than that for men. Dalziel and Massoglia (1956) found that at 60 Hz, 0.5 percent of the women (1 out of 200) could not let go at 6 mA, whereas 0.5 percent of the men could not let go at 9 mA. At 10.5 mA, 50 percent of the women could not let go, whereas at 16 mA, 50 percent of men could not let go. The let-go level where only 0.5 percent of the group tested could let go was 15 mA for women and 23 mA for men. The let-go current threshold can be considered as a "go–no-go" situation. Once immobilized or frozen to a circuit, an individual either will break the contact and live or will not be able to break the contact. If the contact is not broken, then the person's skin contact resistance may decrease and increase the current to the lethal level, ultimately causing death. To break contact and become free of the circuit, the individual must overcome involuntary muscular contraction while enduring the painful shock. Several people have told the author of being on a ladder when they became frozen to a circuit; they freed themselves by kicking the ladder out from under them to allow their body weight to pull them free. This is an example of how the mind still functions although the arm muscles cannot be controlled voluntarily. Other individuals have reported that while frozen to a circuit, they shouted for help, but bystanders said they heard only a whisper.

A value of 6 mA has been determined to be the safe value for let-go current because shocks at this level will not freeze individuals to energized circuits. The ground fault circuit interrupter, described in Chapter 1, detects current unbalances greater than 6 mA between supply and return currents from an appliance and disconnects the circuit in sufficient time to prevent electrocution (UL 943 1985). If a 6 mA current were shocking someone, the individual could still release the contact. Tripping at 6 mA or higher, with the current–time trip characteristic utilized, will ensure that no one will be frozen to the circuit or receive a lethal shock.

Asphyxia

Asphyxia occurs when the passage of continuous current through the chest cavity causes the chest muscles constantly to contract, interfering with breathing (Cabanes 1985). Such currents are below the level that would cause ventricular fibrillation, but are large enough to contract the muscles so that the individual cannot breathe or let go of the circuit. There is some question as to whether people actually die from asphyxia rather than some other mode of death, such as heart damage (Lee 1966).

Respiratory Arrest

Electrical shocks with a current path through the respiratory center can result in respiratory arrest (Lee 1966). The respiratory center in the medulla of the brainstem is at the base of the skull, slightly above a horizontal line from the back of the throat; thus shocks from the head to a limb or between two arms could lead to respiratory arrest.

Studies indicate that shocks at current levels above those that cause ventricular fibrillation can produce respiratory arrest even though the primary path of the current is not through the respiratory center. Shocks in dogs between the foreleg and the hind leg caused convulsions and respiratory arrest at lower current levels than shocks between the two forelegs even though there would be tendency for the current to be greater in the respiratory center in the latter case (Hodgkin, Langworthy, and Kouwenhoven 1973).

Ventricular Fibrillation

The usual mode of death in electrocution is related to interference of the electrical chest current with the normal neural electrical control of heart muscle function. Normally blood is circulated through the body by the pumping action of the contracting heart muscle. When the heart beats, the left ventricle contracts and pumps oxygenated blood through the aorta and arteries in the body. The right ventricle contracts at the same time and pumps oxygenated blood through the pulmonary artery to the lungs for oxygenation. Returning oxygenated blood from the lungs flows through the pulmonary vein into the left atrium, which contracts slightly prior to the ventricular contraction to aid in filling the left ventricle. Unoxygenated blood returns through the venous system and enters the right atrium, which then fills the right ventricle. The contraction of the ventricles when the heart beats provides the pumping action that circulates blood throughout the body. Death will occur if the ventricles do not contract to circulate the blood. Lack of oxygenated blood effects the brain first, causing irreversible brain damage within 3 to 6 minutes.

Ventricular fibrillation is an uncoordinated, asynchronous contraction of the ventricular muscle fibers of the heart in contrast to their normal coordinated and rhythmic contraction. With ventricular fibrillation, the heart seems to quiver rather than to beat, and blood circulation ceases—a truly life-threatening situation. Ventricular fibrillation may be caused by an electrical shock where the path of current is through the chest, such as between two arms, between an arm and a leg, or across the chest. Because blood circulation ceases at the start of ventricular fibrillation, the person becomes unconscious in less than 10 seconds, and irreversible brain damage commences in 3 to 6 minutes unless cardiopulmonary resuscitation is initiated. Cardiopulmonary resuscitation is used as a temporary measure to provide some circulation of oxygenated blood to the brain and heart until a defibrillator can be used; the only way to terminate ventricular fibrillation is to use a defibrillator, which applies a pulse shock to the chest to restore the heart rhythm. Ventricular fibrillation leaves no characteristic evidence for the pathologist after death, but it often is the terminal condition in death from natural causes such as coronary artery blockage or accidental causes such as drowning. This lack of evidence, together with the fact that a person can be electrocuted with no burns on the body, means that an individual can be electrocuted without there being any positive postmortem findings.

A typical electrocardiogram is shown in Figure 2.1, where the time scale is in milliseconds (ms). The P wave, which is the beginning of the heart cycle, indicates atrial contraction. The wave of excitation triggers the QRS complex, which occurs as the ventricles contract as a unit and pump blood. When contraction is completed, repolarization or relaxation of the ventricles takes place during the T phase. Shocks during the T phase are most likely to cause the heart to go into ventricular fibrillation. To defibrillate the heart, it is necessary to employ a counter shock through the chest that is large enough to depolarize all the muscle fibers of the heart, causing all of them to contract simultaneously. When the heart relaxes after this shock, the normal heart cycle may resume unless ventricular fibrillation has continued too long and the heart muscles are not sufficiently oxygenated.

From experiments with animals extrapolated to possible human application, the 60 Hz current value for shocks with the current path through the chest that will produce ventricular fibrillation is given by the following expression:

$$I = \frac{100}{t} \text{ mA rms}$$

where the shock duration, t, is in seconds, and $0.2 \text{ s} < t < 2 \text{ s}$.

For short-duration shocks shorter than a cardiac cycle, the electrical current to cause fibrillation must be large and occur during the vulnerable period, the T wave. Shocks longer than a cardiac cycle can cause premature ventricular contractions that lower the shock threshold current to a minimum after four or five premature ventricular contractions. Using these concepts, a safe current limit has been proposed as 500 mA for shocks less than 0.2 second in duration and 50 mA for shocks longer than 2 seconds (Biegelmeier and Lee 1980). The 500

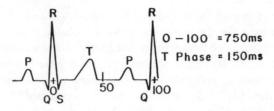

Figure 2.1. Electrocardiogram.

mA shock for less than 0.2 second would have to occur during the T wave of the heart cycle (IEC 479-1 1984).

Asystole

Asystole occurs when the heart is at cardiac standstill and does not beat. Higher current shocks, above about 1 A through the chest cavity, can cause such a condition. Because the higher currents are associated with higher voltage, asystole rather than ventricular fibrillation often is the cause of death for accidents involving 1,000 V or higher-voltage power circuits. Unlike ventricular fibrillation, asystole may convert to a normal heart rhythm in some cases. It is not surprising to find that a lineman may survive asystole with serious burns to the body after a high voltage, high current shock although another person will die of ventricular fibrillation after a low current, 120 V shock without any burn marks.

EFFECT OF FREQUENCY AND WAVEFORM

With electrical shocks from other than power line sources, the specific waveform parameters must be known in order to predict the physiological effects. The waveforms considered below are pulse and impulse, direct current, and alternating currents at frequencies other than 60 Hz.

Pulse and Impulse-Type Shocks

Pulse and impulse-type shocks result from electrical shocks with a time duration much shorter than the heart cycle. Such shocks are provided by capacitor, electrostatic, and lightning discharges.

It has been determined that in an impulse-shock the hazard is related to the electrical energy content in the discharge. These determinations come from animal experiments conducted to develop the defibrillator and from the study of electrical accidents involving capacitor discharge.

A high energy shock can fibrillate the heart at any time during the heart cycle. The lowest-energy shock that can cause fibrillation must occur during the T wave of the heart cycle.

Any pulse shock with an energy content of about 50 J probably is hazardous. Shocks below 0.25 J are disagreeable but probably not hazardous (Dalziel 1971). Underwriters Laboratories, in the standard for electric fence controllers, set an output limit of 0.1 J per pulse for a train of pulses occurring once a second (UL 69 1987). Thus, 0.1 J shocks should not be hazardous. The annoying electrostatic discharge shock produced by walking across a carpet is on the order of 10 mJ. Defibrillators have a maximum output in the range of 200 to 400 J.

Effects of Frequency

For frequencies from 15 to 100 Hz, the current levels for perception, startle reaction, let-go, and ventricular fibrillation are about the same as those for 60 Hz. The effects of frequencies above 100 Hz on the threshold current for perception or startle, let-go current, and current to induce ventricular fibrillation (IEC 479-2 1987) are shown in Table 2.1. The table lists the ratio of the currents necessary at a given frequency to the currents at 60 Hz to produce the same physiological effect. These results demonstrate that from 15 to 1,000 Hz, the current levels for perception or startle and the let-go phenomenon do not change significantly with frequency. From 1 kHz to 10 kHz, a considerably larger current is required for the same effect as the frequency is increased. The current required for ventricular fibrillation increases rapidly at frequencies above 100 Hz.

The direct current shock levels for threshold of perception or startle reaction,

Table 2.1. Ratio of the Current to Produce an Effect at the Indicated Frequency to the Current at 60 Hz to Produce the Same Effect (IEC 479-2 1987).

FREQUENCY (Hz)	THRESHOLD OF PERCEPTION	LET-GO	VENTRICULAR FIBRILLATION
15 to 100	1	1	1
300	1.2	1.15	5
1,000	2.1	1.65	14
3,000	5.0	2.5	*
10,000	12	5	*

*Insufficient data.

about 1.5 to 2 mA, are about three times the current required at 60 Hz. The direct current magnitude for ventricular fibrillation, about 150 mA, is three times that for the 60 Hz shock for shocks longer than 2 seconds, but it is the same (500 mA) as for a 60 Hz shock for shocks shorter than 0.2 second. With direct current shocks there is severe pain when the circuit is made or broken, but there is little pain while the current is maintained. Therefore, there may be no true let-go phenomenon for direct current. Direct current let-go levels were defined as the direct current magnitudes when the test subjects refused to let go because of the anticipated severe jolt (Dalziel and Massoglia 1956).

ELECTRICAL BURNS

Knowledge about the burn characteristics of electrical or lightning injuries can be useful in reconstructing the events that led to an accident. Analysis of the burns received can be helpful in identifying an electrocution, the points of electrical contact, and information about the activities of the victim at the time of the accident. One difficulty posed by accident reconstruction is that an individual can be injured by electricity or even electrocuted without having any external evidence of an electrical burn.

External surface burns on the body can be caused by the heat produced by an arcing in nearby equipment or by the splatter of molten particles from damaged conductors or equipment. A shock from contact with energized equipment can cause external or internal burns. The magnitude of the current, the type of contact, and the duration of the contact determine the severity of the injury.

Burns at the point of skin contact can be caused by arcing to the skin, heating at the point of contact by a high-resistance contact, or large currents. It is possible to have an electrocution with low voltage shocks, such as at 120 V, without any visible marks to the body. Higher-voltage shocks, because of the larger currents and the increased likelihood for arcing to occur, usually will leave contact marks on the skin.

Electrocution can occur while an individual is swimming in water containing electric currents. Electrical currents tend to cause either ventricular fibrillation or continuous muscular contraction so that swimming is impossible and drowning occurs. There usually are no burns marks on the body, and postmortem findings are exactly those for a conventional drowning. A pathologist can determine that death may have been caused by electricity only by having knowledge that there were electrical currents in the water at the time of the apparent drowning.

LIGHTNING

Lightning death and injury may be caused by a direct strike, by currents flowing in the ground, or by a sideflash from a nearby object struck by lightning (Lee

1977). The usual lightning discharge is a constant, unidirectional current source of very high magnitude and short duration. This transient current reaches a peak in about 5 microseconds and is substantially over with 100 microseconds. The peak current has a median value of 30,000 A.

If an individual were struck directly by a lightning discharge, his or her body would in a current path with a peak current of up to 30,000 A or higher. Because of body resistance, a high current will develop a high voltage across the body, producing an arc shunting much of the current out of the body. This will cause burns on the outer body surface. The remaining, large transient current in the body may lead to cardiac asystole or ventricular fibrillation. With asystole, the heart may restart beating after the strike. That is why some people surprisingly have survived a direct lightning strike. The person struck may not have major internal burns because the lightning current, though large, is of very short duration.

When lightning strikes the ground or an object on the ground, there will be lightning currents in the ground near the strike. These currents flowing through the resistance of the earth will develop voltage differences between different points on the surface of the ground. Lightning shocks can occur because of the potential difference along the ground where a person is standing, stepping, or lying down.

A sideflash occurs when an object is struck by lightning and some of the current (sideflash) jumps to a nearby grounded object. For example, a tree struck by lightning will have a substantial voltage to ground along its length because of high internal resistance. This voltage can cause a sideflash of current to a person standing on the ground adjacent to the tree. People standing near a tree struck by lightning often are victims of a sideflash or ground currents flowing out from the tree.

A metal pole with its foundation buried in the ground can develop a large voltage to ground if struck by lightning. The voltage is caused by the ground resistance at the foundation. At the 30,000 A peak, a ground resistance of 10 Ω will have a voltage to ground of 300,000 V, which easily could produce a sideflash to a nearby grounded object.

HIGH VOLTAGE ELECTRIC SHOCKS

Electric shocks at voltages over 600 V can be considered as high voltage shocks. At such voltages there usually is some physical evidence, such as an electrical burn on the body. Because of the higher currents involved, electrocution may be the result of cardiac asystole rather than ventricular fibrillation although ventricular fibrillation can also result. Serious electrical injury is more common than in lower-voltage accidents because the currents are larger, and there is more likelihood of arcing associated with the accident.

An understanding of the physiological effects of electricity, described earlier in this chapter, and knowledge of electrical systems, described in Chapter 1 will help investigators to explain how accidents occurred and why certain injuries were sustained.

Legal Electrocution

Any discussion of the prevention of accidental electrocution may lead to questions about the intentional legal electrocution of criminals. The first legal electrocution

Figure 2.2. Alabama electric chair.

took place at Auburn Prison, New York, on August 6, 1890 (Bernstein 1973). The electrocution systems used today utilize head and calf electrodes. The applied electricity is an alternating current that is cycled between high voltages in the range of approximately 1,500 to 2,500 V and lower voltages in the range of approximately 500 to 1,000 V. Different states have used different sequences for the applications of the voltage. The State of Alabama, for example, has applied 1,800 V for 22 seconds, decreased the voltage to 750 V in 12 seconds, raised the voltage back to 1,800 V in 5 seconds, and then turned off the power. Several such cycles often have been required because the heart continued to beat after such a cycle when the power was turned off. The current at 1,800 V was approximately 7.5 A.

Alternating current is used for electrocutions, as it certainly is more lethal than direct current. Robert Elliott, who executed 387 prisoners while he was the executioner for the State of New York from 1926 to 1939, found that if he applied a high voltage shock sequence of about 2,000 V for 5 seconds and 1,000 V for 30 seconds for two cycles, and then turned off the power, he often would have to reapply power because the prisoner's heart was still beating. If he reduced the voltage from 1,000 V to zero gradually instead of abruptly turning it off at the end of the second cycle, the further application of power would not be necessary (Elliott 1940). Elliott did not explain this phenomenon, but probably at high voltage the heart was in asystole, and it started to beat when the power was removed abruptly. When the voltage was reduced gradually from 1,000 V to zero, the heart went into ventricular fibrillation as the current went through the ventricular fibrillation range, below an ampere. With fibrillation, another cycle would not be necessary. The Alabama cycle probably would tend to defibrillate the heart with the final high voltage pulse. A photograph of the Alabama electric chair is shown in Figure 2.2.

POWER LINE CONTACT ACCIDENTS

The usual power line contact accident case involves contact with either intercity overhead distribution or cross-country transmission lines. These conductors usually are bare (uninsulated) and operate at line-to-ground voltages of 2,400 to 19,900 V for distribution and at higher voltages for transmission systems. Accidents occur with all types of systems and at all voltages, but they are more common with distribution systems where there are miles of lines near inhabited areas. These lines are installed at lower heights than are the higher-voltage transmission lines because of lower code height requirements (*National Electrical Safety Code* 1990). Accidents commonly involve a 7,200 V distribution line, the most widely used primary voltage line.

Crane Power Line Contact

A sketch of a crane on the ground with its boom in contact with a power line is shown in Figure 2.3. The entire crane, being made of metal, will be at the same voltage as the line. Workers on the crane will not be injured as long as they do not leave the crane. All points on the crane are at the same line voltage, and the individuals are like birds on a power line; there is no potential difference between points of contact on the crane. However, a worker standing on the ground and touching the crane can receive a serious shock. This is equivalent to touching the bare power line itself while standing on the ground. A worker on the ground but not in contact with the crane also could receive a shock between contact points on the ground because of a voltage difference produced by the ground currents.

Quite often the overcurrent protection device for the line will not operate to disconnect the power. This is so because the resistance to ground at the vehicle ground contact points is relatively high and limits the ground fault current flow. Even metal outriggers resting on the earth easily can have resistance-to-ground values of several hundred ohms or higher. It usually makes little difference whether the crane or the vehicle has rubber tires no not, as the high voltage will arc either over or through the tires to ground.

After such an accident, it often is claimed that the crane boom was not near the power line, but that the electricity arced a distance of a foot or two from the power line to the boom to cause the mishap. Such an arc is impossible, as a voltage gradient of 75,000 V per inch is necessary for electricity to arc through air. Thus, a 7,200 V line can initiate an arc only over an air gap distance of about 0.1 inch. What probably happens in this case is that the crane boom contacts the power line and initiates an arc as the boom pulls away from the

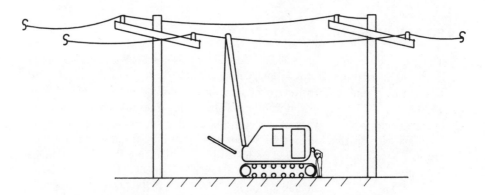

Figure 2.3. Metallic crane boom contact with overhead line.

line. Because only 50 V per inch of voltage gradient is required to maintain the arc, once it is initiated, the 7,200 V can sustain a long arc as the crane boom pulls away. If the boom maintains contact with the line, there may be little damage to the boom and line at the point of contact, as there would be minimum arcing at the low-resistance, metal-to-metal contact point. Additionally, the current would be limited by the usually high ground resistance at the crane–earth contact.

Power Lines on or Near the Earth

Power lines occasionally break because of poor splices, arcing line faults, deteriorated line support structures, or vehicular accidents involving the line supports. A power line on the earth usually will have a high resistance to electrical ground because of the small contact area. The circuit overcurrent protection device for the line may not operate because it may have an operating value of 50 A or higher. In addition, there may be arcing and burning of brush or plant growth at the contact point with the ground. Power lines lying on the ground are particularly hazardous because a person might come in contact with the fallen conductor. One individual who stepped on a downed high voltage conductor fell to the ground from the resulting electric shock, only to be severely burned by ground currents near the point of contact.

A low-hanging power line not touching the ground cannot possibly trip an overcurrent protection device to alert anyone that a problem exists. Occasionally people have walked into low-hanging power lines and have been severely injured or electrocuted. In these cases, too, the line overcurrent protection devices probably did not operate. The fault current flow is limited by the line-to-skin contact impedance and the volumetric body impedance plus the ground contact resistance of the victim.

Tree Contact

In some cases, such as severe weather situations, live power lines become entangled in trees, usually causing minor arcing or burning of the foliage. Generally the circuit overcurrent protection device will not trip because of the relatively high resistance of the tree-to-ground current path, a resistance on the order of thousands of ohms (Defandorf 1956). A child who climbed the tree and touched either the power line or a tree limb in contact with the line could receive a shock, whose severity would depend on the details on the points of contact with the power line. If the grounded neutral line were also in the tree, the shock could be particularly severe, as the resistance path to electrical ground via the neutral line can be relatively small.

Antenna or Metal Pole Line Contact

If an antenna mast or a metal pole resting on the ground contacts a power line, the situation is similar to the crane power line example presented previously. Again, the high resistance of the ground contact point will limit the current so that the circuit overcurrent protection device may not operate. Anyone who now comes in contact with the antenna or metal pole will receive a severe shock. The body of the individual will provide a current path to ground. The voltage difference across the body will be the power line to ground voltage. The ground fault current for the line now will consist of parallel currents through the antenna or pole to ground and the current through the body to ground.

If instead an individual is holding a metal pole above the ground and comes in contact with a power line, the total ground fault current will be smaller than in the previous example. This is true because the only current path to ground will be through the body, and it will be of relatively high resistance.

High Voltage Tester

One type of high voltage insulation tester provides a range of test voltages up to 2,500 V with a current limit of 300 mA. This means that even with a direct short circuit at the output terminals, the current delivered by the device never can exceed the current limit of 300 mA. A worker using this device to test insulation was electrocuted as the worker's wrists came in contact across the output terminals with the output voltage set at 1,750 V. Assuming a minimum hand-to-hand resistance of 500 ohms, the maximum current is seen to be the 300 mA current limit of the device. Because this current is well above the 50 mA threshold for ventricular fibrillation, the worker's death is not surprising.

Pad-Mounted Transformers

Underground distribution systems have pad-mounted transformers housed in enclosures on the ground. The code requires these enclosures to be locked for prevention of entry and possible exposure to the high voltage equipment inside. There are cases where the enclosures have been left unlocked, and children have played inside. Neither they nor responsible adults were aware of the hazard of the exposed primary voltage terminals, fuses, or disconnects.

A Milwaukee child lost most of his arm and suffered massive and serious internal injuries from a high voltage shock in a pad-mounted transformer. His body was in contact with the transformer housing when he touched a high voltage terminal with his hand or arm. The padlock had been reported missing from the transformer door.

Once again, because the high currents involved are well above the ventricular fibrillation range, it is no surprise that electrocution does not occur in many cases, including the above example. A secure locking mechanism, warning signs on the enclosure, and periodic inspections are necessary to prevent such accidents.

LOW VOLTAGE SHOCKS AND INJURY

Low voltage shocks (under 600 V) are more common than high voltage shocks because there are so many more low voltage systems near high population areas. Low voltage shocks tend to cause less tissue damage than high voltage shocks because of the lower currents involved. Ventricular fibrillation can occur, as the body currents resulting from low voltage shocks tend to be in the current range necessary for fibrillation. When there is an arcing fault in a piece of equipment, any individual near the equipment but not necessarily in the current path can receive severe burn injuries from the arc or splatter of metal.

Improper Equipment Grounding

Proper equipment grounding is necessary to ensure plant electrical safety. The purpose of equipment grounding is to ensure that all exposed metal equipment surfaces will be maintained at substantially ground potentials even during a fault condition. A low-resistance internal fault to an exposed metal surface probably will cause the circuit overcurrent protection device to trip if the equipment is properly grounded. Thus, the power to the equipment will be disconnected, removing the hazard. The operation of the overcurrent device occurs because of the relatively low-resistance path provided by the series combination of the internal fault and the equipment ground.

Even in a fault condition of high internal resistance, a properly installed equipment ground will tend to keep exposed metal surfaces at or near ground potentials A fault with a high internal resistance limits the fault current flow to values below that necessary to activate the overcurrent protection device. Thus, even though the equipment power in this situation is is not disconnected, the equipment surfaces still are safe to touch.

With either fault condition, an individual contacting the faulted equipment would not receive a serious shock if there were an equipment ground properly installed. Proper equipment grounding is demonstrated in Figure 2.4b along with the same equipment minus equipment grounding (Figure 2.4a). Without a proper equipment ground, both low- and high-resistance internal faults will energize the exposed metal parts to internal voltage levels. A person contacting the exposed metal can receive a lethal shock in either case. Even in the high-resistance internal

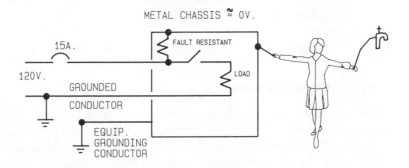

(a) EQUIPMENT WITH INTERNAL FAULT AND WITHOUT
EQUIPMENT GROUNDING. DANGEROUS CONDITION
WITH 120V. TO GROUND ON CHASSIS.

(b) EQUIPMENT WITH INTERNAL FAULT AND WITH
EQUIPMENT GROUNDING. CIRCUIT BREAKER WILL
TRIP IF FAULT RESISTANCE SUFFICIENTLY LOW.
FOR HIGHER FAULT RESISTANCES CHASSIS REMAINS
AT GROUND POTENTIAL.

Figure 2.4. Importance of equipment grounding and internal fault resistance in shock cases.

fault situation, there can sufficient current to be lethal if an individual simultaneously contacts the exposed metal surface of the ungrounded metal chassis and a good electrical ground.

Example: 120 V Applicance Fault

Consider a 120 V appliance with a 50-ohm fault resistance between the 120 V power supply and exposed metal parts. This situation would produce a ground fault current of only 2.4 A if the exposed metal parts were grounded. The usual circuit overcurrent protection device, rated at 15 A or greater, would not usually operate, but there would be no significant hazard because the equipment grounding would limit the voltage between the exposed metal parts and the ground. However, in the same fault condition without equipment grounding, a human body of 500 ohms internal resistance could be exposed to a voltage of 109 V and a potential current of approximately 218 mA. If good body contact were made simultaneously with a ground and the exposed metal parts, a potentially lethal condition would exist.

THE HAZARD IN MIS-WIRING EQUIPMENT WITH THE EQUIPMENT GROUNDING CONDUCTOR INTERCHANGED WITH AN UNGROUNDED CONDUCTOR.

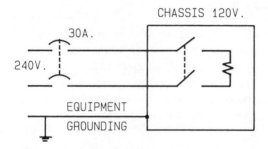

(a) PROPER WIRING FOR A 240V. APPLIANCE

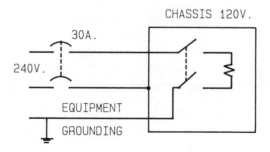

(b) MIS-WIRING 240V. APPLIANCE WITH UNGROUNDED CONDUCTOR CONNECTED TO THE CHASSIS.

Figure 2.5. Importance of having properly wired equipment grounds. (a) Proper wiring for a 240 V appliance. (b) Miswiring of 240 V appliance with ungrounded conductor connected to the chassis.

Sometimes, through either miswiring or incorrect insertion of the power plug into a receptacle, an ungrounded and energized conductor is connected directly to exposed metal parts, as shown in Figure 2.5. This situation is particularly hazardous because the full line voltage is applied to the exposed metal of the chassis. The circuit overcurrent protection device will not operate unless the metal chassis has a good low-resistance connection to the electrical ground. A person touching such equipment could receive a severe shock.

Some older electrical installations have receptacles with only two blades and no third prong for the equipment grounding wire. People using these systems with modern three-pronged appliances frequently remove the third prong or use a "cheater" adapter. The absence of any equipment grounding creates an unsafe electrical environment for users of these systems.

SAFE VOLTAGES

Usually the voltage of a system involved in an electric shock case is known. The higher the voltage, the greater the potential current is in the body of an individual receiving a shock, and the more likely it is that electrocution can occur. A safe voltage, that is, a voltage so low that electrocution is almost impossible, will depend upon the waveforms and the frequency of the applied voltage. Another factor is the source impedance or available current. For example, a 1,000 V power supply with a 100,000-ohm source resistance can deliver a current of only 10 mA or less to someone directly across the output terminals and certainly could not cause an electrocution. Because approximately 50 mA is required for electrocution at 60 Hz, voltages at 60 Hz in the range below about 30 to 50 V are considered safe, with few authenticated cases of electrocution being reported at such voltages (Kouwenhoven 1949). Direct current shocks below 100 V probably are safe, as currents above 150 mA would not be likely.

In article 110-17(a) of the National Electrical Code (1990) there is a discussion of guarding live parts against accidental contact. The requirements for guarding with an enclosure or by location are stated only for systems operating at 50 V or more. This would indicate that at 60 Hz a voltage of 50 V or less is not considered particularly hazardous.

Electric arc welders are limited to a maximum open circuit voltage of 80 V (UL 551 1987), which is the lowest voltage at which satisfactory welding is feasible. Using this reduced voltage, compared to 120 V, will decrease the likelihood of electrocution. Electrocutions have been reported with 80 V at 60 Hz when an operator made good skin contact with the electrodes or had good skin contact with a current path directly across the chest. However, such electrocutions are not common because the voltage is reduced, and any gloves or other clothing in the current path tends to reduce the current below the lethal level.

ELECTRICAL CURRENTS IN WATER

Electrocutions can occur when an individual is in water and electric currents are present (Dalziel 1966). For example, children have been electrocuted in bathtubs when an appliance, such as a hair dryer or electric heater, has fallen into the tub. In other instances, swimmers in a lake or a pool have been shocked by electric currents that have been injected into the water by defective electrical equipment or systems.

An energized conductor that is exposed to a pool or a body of water can produce currents in the water (Smoot and Bentel 1964). The current's magnitude will depend upon the shape and the size of the conductor–water contact surface, the conductivity of the water, and the resistance in the current path to ground (Novotny and Priegel 1974). The energized conductor in contact with the water can be analyzed much like a grounding electrode with a fairly high resistance to ground because of the relatively small exposed contact surface. This would tend to limit the current so that the circuit overcurrent protection device might not operate.

Consider the situation of a 120 V electric appliance, such as a hair dryer or a space heater, being dropped into a bathtub full of water. There will be a current path between any energized portions of the appliance and grounded plumbing, which are both in contact with the water. There can even be current in the water with the appliance switched off if the energized switch terminal is in contact with the water. The currents in the water can be sufficient to electrocute a child or to cause the child to drown because of immobilization of the muscles, which would prevent the child from getting out of the water. The subject's grasping the appliance in the water could be particularly hazardous if another part of the body were in contact with the grounded plumbing. A ground fault circuit interrupter would prevent such electrocutions, as it would deenergize the appliance as soon as it fell into the water if there were a path to ground. An appliance with equipment grounding would be unlikely to cause an electrocution because the grounded exposed metal parts would tend to reduce the current flow in the water to the grounded plumbing.

In the ocean, a lake, or a pool in the earth, energized conductors in contact with the water will produce currents between the conductor and the earth or grounded electrical surfaces in contact with the water. Once again, the currents depending on the resistance of the ground path may not be large enough to operate the overcurrent protection. One such situation occurred when a defective submersible pump, lacking an equipment ground, had a fault that energized the pump case. Another case involved an all-metal boat with a miswired shore electrical power system and a metal hull that was energized to 120 V above ground. People on the boat did not notice any problem because all points of potential body contact on the boat were at the same 120 V potential. When they boarded (stepped on) the boat, they did not receive a shock because the pier was made of wood. However, currents entering into the water from the metal hull killed fish under and near the boat; and a swimmer drowned after entering into the electric field and suffering sustained muscle contractions due to the resulting shock.

Electric Worm Probe

An electric device that has been used for many years to harvest earthworms consists of a metallic electrical probe about the size of a large knitting needle. The probe is inserted into the ground and connected to an ungrounded 120 V conductor from a branch circuit that energizes the probe to 120 V. Current is injected into the ground in a current path from the probe to the electrical ground through the earth. The magnitude of the current will be limited by the high ground resistance at the probe-to-earth interface so that a 15 or 20 A branch circuit breaker will not open. The actual current would be in the range of 1 or 2 A. These currents in the ground seem to annoy earthworms and bring them to the surface.

Worm probes are dangerous and should not be used. The probe represents a bare 120 V conductor lying on the ground. If a person contacts the probe and the moist surface of the earth, a serious shock can result. An even worse situation may occur if contact is made with the probe and an electrically grounded object such as a telephone pedestal, a pad-mounted transformer enclosure, or an electrical conduit. Such contact is potentially lethal.

REFERENCES

ANSI C101.1. 1986. *Leakage Current for Appliances*. New York: American National Standards Institute.
Bernstein, T. 1973. "A grand success." The first legal electrocution was fraught with controversy which flared between Edison and Westinghouse. *IEEE Spectrum* 10(2): 54–58.
Bernstein, T. 1983. Electrocutions and fires involving 120/240 V appliances. *IEEE Transactions on Industry Applications*. IA-19(2): 155–59.
Biegelmeier, G. 1985. New knowledge on the impedance of the human body. In *Electrical Shock Safety Criteria*, ed. J. E. Bridges, G. L. Ford, I. A. Sherman, and M. Vainberg, pp. 115–132. New York: Pergamon Press.
Biegelmeier, G. and W. R. Lee. 1980. New considerations on the threshold of ventricular fibrillation for AC shocks at 50–60 Hz. *Proceedings IEEE* 127(2): 103–110.
Cabanes, J. 1985. Physiological effects of electric currents on living organisms, more particularly

humans. In *Electrical Shock Safety Criteria,* ed. J. E. Bridges, G. L. Ford, I. A. Sherman, and M. Vainberg, pp. 7–24. New York: Pergamon Press.

Dalziel, C. F. 1966. Electric shock hazards of fresh water swimming pools. *IEEE Transactions on Industry and General Applications* IGA-2(4): 263–273.

Dalziel, C. F. 1971. Deleterious effect of electric shock. In *Handbook of Laboratory Safety,* 2nd edition, ed. N. V. Steere, pp. 521–527. Cleveland, OH: Chemical Rubber Co.

Dalziel, C. F. and F. P. Massoglia. 1956. Let-go currents and voltages. *AIEE Transactions* 75 Part II: 49–56.

Defandorf, F. M. 1956. Electrical resistance to the earth of a live tree. *IEEE Transactions Power Apparatus and Systems* PAS 56: 936–941.

Elliott, R. G. with A. R. Beatty. 1940. *Agent of Death, the Memoirs of an Executioner.* New York: E.P. Dutton.

Geddes, L. A. and L. E. Baker, 1967. The specific resistance of biological material—a compendium of data for the biomedical engineer and physiologist. *Medical and Biological Engineering* 5: 271–293.

Hodgkin, B. C., O. Langworthy, and W. B. Kouwenhoven. 1973. Effect on breathing *IEEE Transactions Power Apparatus and Systems* PAS-92(4): 1388–1391.

IEC 479-1. Second Edition. 1984. *Effects of Current Passing through the Human Body.* Part 1: General Aspects. Geneva: International Electrotechnical Commission.

IEC 479-2. Second Edition. 1987. *Effects of Current Passing through the Human Body.* Part 2: Special Aspects. Geneva: International Electrotechnical Commission.

Kahn, F. and L. Murray. 1966. Shock free electric appliances. *IEEE Transactions on Industry and General Applications* IGA-2(4): 322–327.

Kouwenhoven, W. B. 1949. Effects of electricity on the human body. *Electrical Engineering* 68(3): 199–203.

Lee, W. R. 1966. Death from electric shock. *Proceedings IEEE* 113(1): 144–148.

Lee, W. R. 1977. Lightning injuries and death. In *Lightning,* Vol. 2, ed. R. H. Golde, pp. 521–543. New York: Academic Press.

National Electrical Code 1990. NFPA 70 1990 Edition. Quincy, MA: National Fire Protection Association.

National Electrical Safety Code. ANSI C2-1990. New York: Institute of Electrical and Electronics Engineers.

Novotny, D. W. and G. R. Priegel. 1974. *Electrofishing Boats.* Technical Bulletin No. 73. Madison, WI: State of Wisconsin Department of Natural Resources.

Smoot, A. and J. Stevenson. 1968. Subject 965-1. Report on investigation of reaction currents. Melville, NY: Underwriters Laboratories.

Smoot, A. W. and C. A. Bentel. 1964. Electric shock hazard of underwater swimming pool light fixtures. *IEEE Transactions on Power Apparatus and Systems* 83: 945–964.

UL 69. 1987. *Standard for Safety. Electric-Fence Controller.* Northbrook, IL: Underwriters Laboratories Inc.

UL 551. 1987. *Standard for Safety. Transformer-Type-Arc-Welding Machines.* Northbrook, IL: Underwriters Laboratories Inc.

UL 943. 1985. *Standard for Safety, Ground-Fault Circuit Interrupters.* Northbrook, IL: Underwriters Laboratories Inc.

3

Selecting the Proper Size for Conductors and Overcurrent Protection Devices

Gregory Bierals

Most electricity-related industrial accidents result from an electrical system's inability to carry a continuous load without excessive heat buildup or to handle safely a short circuit or ground fault condition when it occurs. Both of these potentially hazardous conditions can be minimized by properly sizing the conductors and the conductors' overcurrent protection device.

In selecting the proper size for a conductor and a protective device it is important to understand the appropriate factors that may apply and how the conditions of application relate to the National Electrical Code (NEC) (*National Electrical Code* 1990), the electrical code that is legally applicable throughout most of the United States.

This analysis begins with an examination of the scope of Article 240-1 of the NEC and the fine-print note that follows. This fine-print note indicates that overcurrent protection for conductors and equipment is provided to open the circuit if the current reaches a value that will cause an excessive or dangerous temperature in conductors or insulation. The requirement to protect conductors at their listed ampacities is mandated in Article 240-3. The protection of flexible cords, extension cords, and fixture wires is referenced in Article 240-4.

FACTORS THAT DETERMINE THE PROPER WIRE SIZE FOR A CONDUCTOR

The conductor ampacities listed in NEC Tables 310-16 through 310-19 are based on four principal determining factors related to conductor temperature. These factors, which are referenced in Article 310-10, include:

1. The ambient temperature, which may vary along the conductor length as well as from time to time.
2. Heat generated internally in the conductor as the result of load current flow.
3. The rate of heat dissipation into the ambient medium surrounding the conductor.
4. The effect of adjacent load-carrying conductors, also called the proximity effect.

For example, examination of the listed ampacity of a #3 THHN copper conductor in Table 310-16 shows that the conductor is rated at 110 A. This value corresponds to the second of the four principal determinants of conductor operating temperature. THHN is listed in the 90°C operating-temperature column of Table 310-16. Further examination of the other three determining factors shows that the table lists an ambient temperature that is not to exceed 30°C (86°F). Should the ambient temperature be exceeded, appropriate correction factors are specified to reduce the conductor ampacity and its operating temperature. This table specifies that the #3 THHN conductor is installed in a raceway surrounded by free air (this would also include cable types AC, NM, NMC, and SE). Of course, thermal insulation or any other restricting medium certainly would affect the conductor operating temperature. A revision in the 1990 NEC makes Table 310-16 apply also to buried systems.

Table 310-16 is applicable only when there are not more than three current-carrying conductors in the raceway or the cable. A neutral conductor that carries only the unbalanced current from the other conductors, as in normally balanced circuits of three or more conductors, does not count as a current-carrying conductor (see Note 10a for Tables 310-16 through 310-19). However, a neutral conductor that is part of a three-phase, four-wire, wye connected system supplying electric discharge lighting or data processing systems does count as a current-carrying wire because of harmonic currents that are present in the neutral, even under balanced load conditions. (See Note 10C, Tables 310-16 through 310-19.)

Derating (adjustment) factors are presented in a table in Note 8a following the ampacity charts, to be used in reducing the ampacity rating when a raceway or a cable contains more than three current-carrying conductors. This derating

Table 3.1. Adjustment Factors for Adjacent Conductors (assuming load diversity of 50%).

# OF WIRES	CORRECTION FACTOR
4–6 conductors	80%
7–9 conductors	70%
10–24 conductors	70%
25–42 conductors	60%
43 or more conductors	50%

requirement adjusts for the fourth principal factor that determines a conductor operating temperature, the heating effect of adjacent load-carrying conductors. Based on these specified conditions of use, a #3 THHN copper wire will not exceed an operating temperature of 90°C.

A footnote to the table, presented in Note 8a, indicates that adjustment factors for the heating effects of adjacent load-carrying conductors are based on the assumption of a load diversity of 50 percent for the conductors. These adjustment factors are presented in Table 3.1.

According to Note 8a, these correction factors are valid, starting with ten or more wires, if no more than 50 percent of the total number of conductors are loaded at any one time. If more than 50 percent are loaded simultaneously, then the derating factors specified in Table 3.1 need to be reduced still further.

In this case the adjustment factors presented in Table 3.2 should be used. These derating factors assume no diversity in the loads served by the conductors. Note that the derating factors are the same in both tables for four to nine conductors.

Table 3.2. Adjustment Factors for Adjacent Conductors (assuming no load diversity).

# OF WIRES	CORRECTION FACTOR
4–6 conductors	80%
7–9 conductors	70%
10–20 conductors	50%
12–30 conductors	45%
31–40 conductors	40%
41–60 conductors	35%

Example: Adjustment Factors for Ampacity Calculations

If there were a group of 20 current-carrying wires in a raceway, the original correction factor would have been 70 percent. If these wires had a normal ampacity of 20 A, after application of the correction factor the new ampacity would be 14 A (20 × 0.70 = 14). If the watts loss of these wires were examined hypothetically, assuming a resistance for each conductor of one ohm, this loss would be calculated as:

$$\text{Watts loss} = I^2R = 14^2 \times 1 = 196 \times 1 \text{ ohm} = 196 \text{ watts loss}$$

per conductor. Then, 196 watts loss per conductor × 10 conductors (50 percent of the total) equals a total of 1,960 W. This would then be the maximum watts loss permitted by the footnote. If all 20 wires were to be loaded, the load should be arranged so as not to exceed a total watts loss of 1,960. Therefore, each individual wire could not exceed a total loss of 98 W (1,960 W divided by 20 wires equals 98 W per wire).

Now solving for the current I by taking the square root of 98 W ($I^2 \times R$), one would determine that the total ampere load and ampacity of each wire should not exceed 9.9 A, or 10 A ($\sqrt{[I^2 \times 1]} = \sqrt{I^2} = I = \sqrt{98} \text{ W} = \sqrt{98} = 9.9$).

The second listed correction factor for 20 wires is 50 percent (Table 3.2). Applying this value to the original 20 A rating yields 10 A:

> 20 A (original ampacity rating)
> × .50 (factor when more than 50% of wires are loaded)
> 10 A (new ampere rating of conductor)

In situations where two or more correction factors apply, one should multiply all of the factors together to obtain an overall derating factor.

For instance, consider a situation where both excessive ambient temperature and the proximity effect are present:

> 0.96 (correction factor from Table 310-16 for
>   THHN-copper in 86°F to 95°F ambient)
> × .80 (factor for 4 to 6 current-carrying wires in raceway)
> 0.768 (total derating factor for both conditions)

The final derating factor would be applied to the original conductor ampacity. *Note.* According to the NEC rules, if a computation of conductor ampacity results in a fraction of an ampere of less than .5, such fraction may be dropped from the calculation. If the fractional value is .5 or more, one uses the next highest ampere. (See Section B, Chapter 9 of the NEC.)

FACTORS THAT DETERMINE THE PROPER SIZE OF OVERCURRENT PROTECTION DEVICES

According to the fine-print note that follows Article 240-1, overcurrent protection for conductors and equipment is provided to open the circuit if the current reaches a value that will cause an excessive or dangerous temperature in conductors or conductor insulation. Other conditions also must be considered with relation to conductor protection. Article 240-3 requires that conductors, other than flexible cords and fixture wires, be protected at their ampacities. There are eight exceptions following this reference, and many are in common use.

Exception 1 allows remote control circuit conductors to be protected by overcurrent devices rated above their ampacities. (See Table 725-31a and Figure 3.1.)

Exception 2 permits conductors of a two-wire single-voltage secondary of a transformer to be protected by properly sized primary protective devices, provided this primary protection is in agreement with Article 450-3 and does not exceed the value determined by multiplying the secondary conductor ampacity by the secondary-to-primary transformer voltage ratio. If the transformer had more than two secondary conductors, for example, single-phase three-wire or three-phase, then secondary conductor protection would be required, with the possible exception of a 10-foot tap rule condition. (See Figure 3.2.)

Example: Exception 1

Power Limited Class 2 Circuit

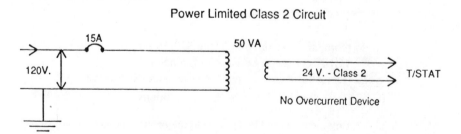

Figure 3.1. Exception 1, power limited Class 2 circuit.

Example: Exception 2

25KVA

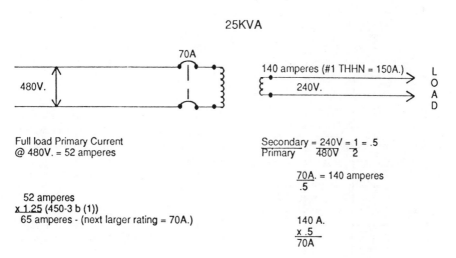

Full load Primary Current
@ 480V. = 52 amperes

$$\frac{Secondary}{Primary} = \frac{240V}{480V} = \frac{1}{2} = .5$$

$$\frac{70A.}{.5} = 140\ amperes$$

52 amperes
x 1.25 (450-3 b (1))
65 amperes - (next larger rating = 70A.)

140 A.
x .5
70A

Figure 3.2. Exception 2, two-wire single-voltage secondary.

Exception 3 permits conductor protection to be omitted where the interruption of the circuit would create a hazard. An example used here is a material-handling magnet circuit. The inference is that short-circuit protection is required, based on the short-circuit withstand rating of the conductor. (See Figure 3.3.)

Exception 4 normally would allow a conductor with an ampacity that does not correspond with a standard rating of a protective device (Article 240-6) to be protected with the next higher standard rating of a protective device, provided that this next larger rating does not exceed 800 A and the conductor is not part of a multi-outlet branch circuit supplying receptacles for cord and plug–connected portable loads. (See Figure 3.4.)

Exception 5 would permit tap conductors to be protected at values in excess of their ampacities. (See Articles 210-19c and 240-21 Exceptions 2, 3, 5, 8, 9, and 10, and 364-10 and 11 and Part D of Article 430.) (See Figure 3.5.)

Exception 6 would permit motor and motor control circuit conductors to be protected at values above their ampacities. Because of inrush current conditions, overcurrent devices typically are sized larger than the motor and conductor rating as referenced in Articles 430-52 and 440-22 and Table 430-152. Of course, most motors are considered to be continuous loads (see Article 430-33), and Article 430-22a would require that the motor full load current be increased by a value of 125 percent. In addition, the motor full load current rating would be determined from Tables 430-147, 148, 149, or 150, as opposed to the motor nameplate

Example: Exception 3

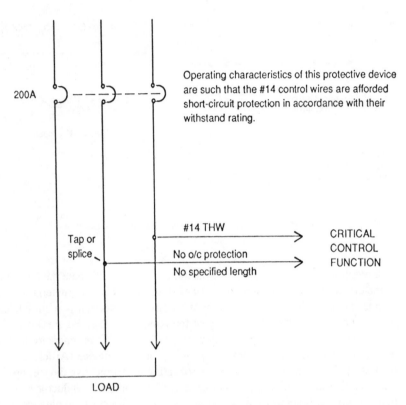

Figure 3.3. Exception 3, when interruption of circuit would create a hazard.

current rating, in agreement with Article 430-6a. For motor control conductors, Table 430-72b permits a protective device to be set at a value above the wire ampacity according to Columns B and C of this table. (See Figure 3.6.)

Exception 7 recognizes that capacitor circuit conductors may be protected at values above their ampacities (see Article 460-8a and b). A separate overcurrent device is not required for a capacitor connected on the load side of a motor protective device. Generally, the ampacity of capacitor circuit conductors is required to be at least 135 percent of the rated current of the capacitor and at least one-third the ampacity of the motor circuit conductors. (See Figure 3.7.)

Example: Exception 4

Note: No derating factores have been applied to the conductor ampacity.

Figure 3.4. Exception 4, use of next highest standard protective device.

Example: Exception 5 - Article 240-21 exception 3 - 25' tap

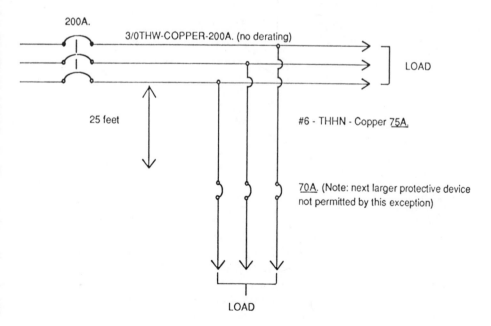

Figure 3.5. Exception 5, 25-foot tap rule.

Example: Exception 6

460 Volts - 3 phase (only one phase shown)

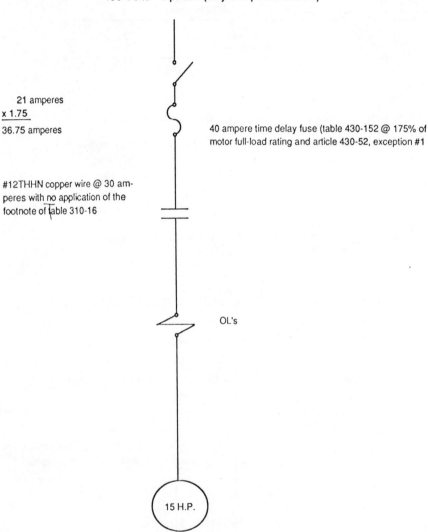

21 amperes (Table 430-150 code letter "F")

Figure 3.6. Exception 6, sizing to accommodate in-rush current.

Example: Exception 7

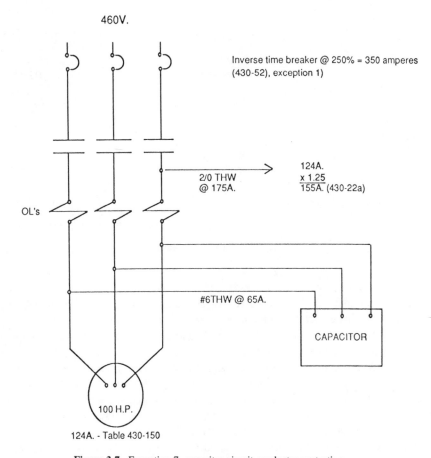

Figure 3.7. Exception 7, capacitor circuit conductor protection.

Exception 8 deals with welder circuits, which comply with Articles 630-11 and 630-12. Article 630-12a, for example, would permit a protective device rated or set at not more than 200 percent of the rated primary current of the welder. Article 630-11a would allow the supply conductors for an individual welder to be determined by multiplying the rated primary current in amperes given on the welder nameplate by a specific factor based on the duty cycle or time rating of the welder specified in Article 630-11a. For example, for a duty cycle of 80, the multiplier would be .89 (see Article 630-11b for a group of welders). (See Figure 3.8.)

Example: Exception 8

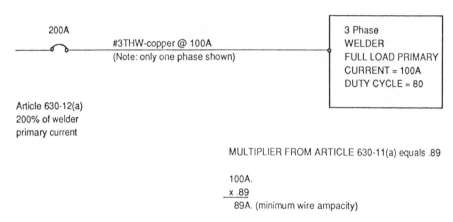

Figure 3.8. Exception 8, welder circuits.

The footnotes of Tables 310-16, 17, 18, and 19 have been modified in the NEC to indicate that the references of the overcurrent protective devices for #14, #12, and #10 copper—15 A, 20 A, and 30 A, respectively—do not apply to the appropriate exceptions of Section 240-3.

An additional note regarding the example of Exception 2 concerns the listing instruction from the UL *Electrical Construction Materials Directory* (Green Book). The Green Book indicates that the terminal provisions of the 70 A circuit breaker shown in this drawing would be relative to the 60°C temperature rating of the conductor attached to these terminals unless otherwise specified. Therefore, the maximum load that could be applied to the #6 THHN cooper conductor is the respective load current rating of a #6 copper wire at 60°C, or 55 A. If an applied load exceeded this value, then the listing instruction in the UL *Directory* would not be satisfied, and Article 110-3b would be in violation. Also, if a circuit breaker rating exceeds 100 A, the temperature rating of the terminal is relative to the 75°C load current rating of the wire that is attached to this terminal. Also one should check the terminal temperature rating of the equipment being supplied.

INTERRUPTING RATINGS OF EQUIPMENT

The subject of interrupting ratings is covered in Section 110-9 of the NEC, which states that "equipment intended to break current at fault levels shall have an interrupting rating sufficient for the system voltage and the current which is available at the line terminals of the equipment." It also states that "equipment intended to break current at other than fault levels shall have an interrupting rating at system voltage sufficient for the current that must be interrupted." An example of the reference to current at other than fault levels is the locked rotor interrupting ratings of enclosed switches with appropriate "horsepower ratings" for a typical motor. Also consideration must be given to the ratings of contactors, motor starters, and circuit breakers for connecting or disconnecting loads at operating current levels that could include a condition of transformer inrush current. Sections 430-109, 430-83, 430-110c(1), 440-12a(2), and 440-12b(1) address the interrupting ratings of switches and circuit breakers used with motors and air conditioning equipment.

The interrupting rating at "fault levels" is the short-circuit interrupting rating of fuses and circuit breakers, listed in symmetrical amperes, in agreement with published data. Some people use "interrupting rating" and "interrupting capacity" interchangeably; however, there is a distinct difference between these two terms. According to IEEE definitions, the meaning of "interrupting rating" is the highest current at rated voltage that an overcurrent protective device is intended to interrupt under specific test conditions. "Interrupting capacity" is the highest current at rated voltage that the device can interrupt. It is the interrupting rating that is listed in published data, and this value is in short-circuit symmetrical amperes (rms). A good design practice would be to consider the interrupting capacity of the protective device as its interrupting rating because this is the actual current the device interrupts during testing.

A method to calculate the short-circuit amps (sca) that can be delivered at the secondary terminals of a transformer, assuming an infinite, unlimited, short-circuit current available at the primary, uses the following formula:

$$I\text{sca} = (\text{Full Load Amperes}) \times \frac{(100)}{(\%Z)}$$

where %Z is the nameplate impedance of the transformer

Example.   Given a 500 kVA, three-phase, 2.0% impedance, 480 V secondary transformer, find the short-circuit current available at the secondary terminals.

Solution.   Either by calculation or from tables such as Table 3.3 and Table 3.4, the full load current is found to be 601 A. Substitution into the above formula gives:

$$I\text{sca} = \frac{601 \text{ A} \times 100}{2\%} = 30{,}050 \text{ sca (rms)}$$

Table 3.3. Three-Phase Transformer, Full-Load Current Rating (in amperes).

VOLTAGE (LINE-TO LINE)	TRANSFORMER kVA RATING								
	150	167	225	300	500	750	1000	1500	2000
208	417	464	625	834	1388	2080	2776	4164	5552
220	394	439	592	788	1315	1970	2630	3940	5260
240	362	402	542	722	1203	1804	2406	3609	4812
440	197	219	296	394	657	985	1315	1970	2630
460	189	209	284	378	630	945	1260	1890	2520
480	181	201	271	361	601	902	1203	1804	2406
600	144	161	216	289	481	722	962	1444	1924

UL Standard 1561 allows the marked nameplate impedance of a transformer to vary by plus or minus 10 percent. Therefore, to provide Isca for the worst-case situation of the example, a modified calculation for Isca is used:

$$I\text{sca} = \frac{601 \times 100}{2\% \times 0.90} = 33{,}389 \text{ sca (rms)}$$

In the example, the protective devices for the transformer secondary would be required to have an interrupting rating of at least 33,389 sca (rms). In addition, the total impedance of all connected circuits downstream of these protective devices would have to be known so that the fuses or circuit breakers throughout the system could be rated properly. If the available fault exceeds the interrupting rating of a fuse or a circuit breaker, the device can be damaged or destroyed. One must keep in mind that an overcurrent protective device with a high interrupting rating does not necessarily ensure total component protection. Wire,

Table 3.4. Single-Phase Transformer, Full-Load Current Rating (in amperes).

VOLTAGE	TRANSFORMER kVA RATING									
	25	50	75	100	150	167	200	250	333	500
115/230	109	217	326	435	652	726	870	1087	1448	2174
120/240	104	208	313	416	625	696	833	1042	1388	2083
230/460	54	109	163	217	326	363	435	544	724	1087
240/480	52	104	156	208	313	348	416	521	694	1042

cables, panelboards, motor starters, lighting ballasts, and other equipment must be provided with short-circuit protection in agreement with their withstand capabilities.

So, as will be demonstrated in the next section, it is necessary to protect conductors and equipment with regard to specific, listed ampacities (NEC section 240-3), but one also must consider other circuit characteristics.

CIRCUIT IMPEDANCE AND OTHER CHARACTERISTICS

Examination of NEC Section 110-10 shows some very important considerations relating to conductor and component protection. Unfortunately, this very important aspect of component protection often is overlooked. Section 110-10 indicates that "the overcurrent protective devices, the total impedance, the component short-circuit withstand ratings, and other characteristics of the circuit to be protected shall be so selected and coordinated as to permit the circuit protective devices used to clear a fault without the occurrence of extensive damage to the electrical components of the circuit." This "fault" could be the result of either a short-circuit or a ground fault. The components of an electrical system, including wire, distribution equipment, busways, switches, motor starters, and lighting ballasts, have limited short-circuit withstand ratings. For instance, the reader can obtain a copy of the Insulated Cable Engineers Association data on the "Short-Circuit Characteristics of Insulated Cable" (Publication P-32-382, available from ICEA, Inc., P.O. Box P, South Yarmouth, MA 02664). This booklet lists the allowable short-circuit currents for various insulated copper and aluminum conductors, sizes from #10 AWG to 1,000 MCM (kcmil) based on one-cycle through 100-cycle ratings.

Consideration of the operating characteristics of a conductor's protective device (e.g., fault-clearing time, let-thru current, etc.) determines whether the wire is properly protected. In addition, the listing instructions included in the UL *Electrical Construction Materials Directory* (Green Book), under the headings of Panelboards, Enclosed Switches, Motor Controllers, Motor Control Centers, and Busways, specify that equipment be marked with their maximum withstand ratings in rms symmetrical amperes. Certainly the rupturing stresses of short-circuit currents could damage or completely destroy electrical equipment because of the effects of the high-strength magnetic fields that surround conductors and components and the tremendous heat energy associated with fault conditions. A suggestion for determining the available fault currents throughout an electrical system is to calculate these fault currents using a point-to-point method. IEEE Standard 141-1976 (Red Book) is an excellent reference for calculating short-circuit currents at each point in a distribution system.

PROCEDURE FOR POINT-TO-POINT SHORT-CIRCUIT CALCULATION

Example Calculation

The following is an example of a typical point-to-point calculation procedure for the circuit presented in Figure 3.9.

1. As above, either by calculation or by referencing Table 3.3, determine the transformer full-load amperes:

$$I = \frac{(KVA \times 1,000)}{1.73 \times voltage} = \frac{500 \times 1,000}{1.73 \times 208} = 1,388 \text{ A}$$

2. Determine the worst-case short-circuit amperes (rms) at the transformer secondary by reducing the transformer nameplate impedance by 10%:

$$I\text{sca} = \frac{1,388 \text{ A} \times 100}{2\% \times 0.90} = 77,173 \text{ sca (rms)}$$

3. Calculate an F factor for three-phase faults as defined by the following formula:

$$F = \frac{1.73 \times L \times I\text{sca}}{C \times V \text{ (line-to-line)}}$$

where L equals the length of the circuit (in feet) to the fault point, C equals an appropriate constant, obtained from Table 3.5 or Table 3.6, and Isca equals the available short-circuit current in amps. Substituting circuit and table values into the equation yields:

$$F = \frac{1.73 \times 25 \times 77,173}{20,000 \times 4 \times 208} = 0.2006$$

Note. In calculating the F factor in step 3, one should refer to the tables of C values (Tables 3.5 and 3.6) and select the appropriate value. For parallel feeders (as in the example problem), one should multiply the C value by the number of parallel conductors per phase (factor 4 in the above equation).

For single-phase center tapped transformers, the calculations would be based on the following formulas:

Single phase line-to-line fault:

$$F = \frac{2 \times L \times I\text{sca}}{C \times \text{line-to-line voltage}}$$

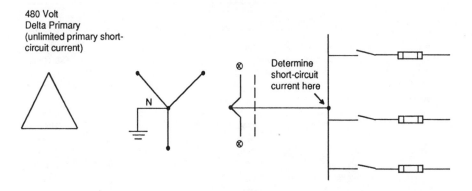

208V/120
Wye secondary

480 Volt
Delta Primary
(unlimited primary short-
circuit current)

Determine
short-circuit
current here

N

500 KVA
3 phase - 2% Z

25 feet - 4 - 500KcMil conductors in parallel
per phase in steel conduit
⊗ to other services

Figure 3.9. Example circuit for short-circuit calculation.

Table 3.5. *C* Values for Busway.

| | PLUG-IN BUSWAY | | FEEDER BUSWAY | | HIGH-IMPED. BUSWAY |
AMPACITY	COPPER	ALUMINUM	COPPER	ALUMINUM	COPPER
225	28700	2300	18700	12000	—
400	38900	34700	23900	21300	—
600	41000	38300	36500	31300	—
800	46100	57500	49300	44100	—
1000	69400	89300	62900	56200	15600
1200	94300	97100	76900	69900	16100
1350	119000	104200	90100	84000	17500
1600	129900	120500	101000	90900	19200
2000	142900	135100	134200	125000	20400
2500	143800	156300	180500	166700	21700
3000	144900	175400	204100	188700	23800
4000	—	—	277800	256400	—

Reproduced through the courtesy of the Bussman Company, St. Louis, MO.

Table 3.6a. *C* Values for Conductors.

	COPPER THREE SINGLE CONDUCTORS			
	STEEL CONDUIT		NONMAGNETIC CONDUIT	
AWG OR MCM	600 V and 5 kV NONSHIELDED	5 kV SHIELDED AND 15 kV	600 V AND 5 kV NONSHIELDED	5 kV SHIELDED AND 15 kV
12	588	—	588	—
10	909	—	909	—
8	1429	1230	1429	1230
6	2222	1940	2273	1949
4	3333	3040	3448	3070
3	4167	3830	4348	3870
2	5000	4670	5263	4780
1	6250	5750	6250	5920
1/0	7692	6990	7692	7250
2/0	9091	8260	9091	8770
3/0	10638	9900	11364	10700
4/0	12500	10800	13514	12600
250	13699	12500	16949	14000
300	15385	13600	17857	15500
350	16667	14700	18868	17000
400	17857	15200	20408	17900
500	20000	16500	23256	19700
600	21277	17200	25000	20900
750	23256	—	31250	—
1000	25000	—	31250	—

Reproduced through the courtesy of the Bussman Company, St. Louis, MO.

Single phase line-to-neutral fault:

$$F = \frac{2 \times L \times I\text{sca}}{C \times \text{line-to-neutral voltage}}$$

where L, C, and Isca are as defined previously.

4. Determine an M (multiplier) value either by making a calculation using the value for F or by making a selection from Table 3.7:

$$M = \frac{1}{1 + F}$$

$$M = \frac{1}{1 + 0.2006} = 0.8329$$

Table 3.6b. *C* Values for Conductors.

| | COPPER THREE SINGLE CONDUCTORS | | | |
| | STEEL CONDUIT | | NONMAGNETIC CONDUIT | |
AWG OR MCM	600 V and 5 kV NONSHIELDED	5 kV SHIELDED AND 15 kV	600 V AND 5 kV NONSHIELDED	5 kV SHIELDED AND 15 kV
12	—	—	357	357
10	—	—	555	556
8	1230	1230	909	909
6	1950	1950	1388	1408
4	3080	3090	2173	2173
3	3880	3900	2702	2702
2	4830	4850	3333	3333
1	6020	6100	4000	4166
1/0	7410	7580	5000	5000
2/0	9090	9350	6250	6250
3/0	11100	11900	7142	7692
4/0	13400	14000	9090	9090
250	14900	15800	10000	10638
300	16700	17900	11363	12195
350	18600	20300	12500	13698
400	19500	21100	13698	15384
500	21900	24000	15625	17543
600	23300	25700	17241	19607
750	25600	28200	19230	22222
1000	—	—	21739	25641

Reproduced through the courtesy of the Bussman Company, St. Louis, MO.

5. Calculate the available short-circuit current (symmetrical) at the point of the fault:

$$I\text{sca (at fault)} = I\text{sca (at beginning of circuit)} \times M$$
$$= 77{,}173 \times 0.8329 = 64{,}277 \text{ sca}$$

As can be seen by application of this point-to-point calculation, there is 77,173 A (rms) of short-circuit current available at the secondary terminals of the transformer. At a point approximately 25 feet downstream at the location of the main service panel, a value of 64,277 A (rms) available short-circuit current is calculated. The withstand rating of the distribution equipment would have to be listed to withstand 64,277 or more amperes of fault current. If it were not, the installation of this equipment would not be in compliance with Article 110-10. The installation of a protective device with proper withstand capabilities and

Table 3.7. M (multiplier) Values.

F	M	F	M
0.01	0.99	1.50	0.40
0.02	0.98	1.75	0.36
0.03	0.97	2.00	0.33
0.04	0.96	2.50	0.29
0.05	0.95	3.00	0.25
0.06	0.94	3.50	0.22
0.07	0.93	4.00	0.20
0.08	0.93	5.00	0.17
0.09	0.92	6.00	0.14
0.10	0.91	7.00	0.13
0.15	0.87	8.00	0.11
0.20	0.83	9.00	0.10
0.25	0.80	10.00	0.09
0.30	0.77	15.00	0.06
0.35	0.74	20.00	0.05
0.40	0.71	30.00	0.03
0.50	0.67	40.00	0.02
0.60	0.63	50.00	0.02
0.70	0.59	60.00	0.02
0.80	0.55	70.00	0.01
0.90	0.53	80.00	0.01
1.00	0.50	90.00	0.01
1.20	0.45	100.00	0.01

Reproduced through the courtesy of the Bussman Company, St. Louis, MO.

current-limiting features ahead of this equipment would then be necessary to achieve compliance.

In an actual case it would be necessary to continue to calculate the available short-circuit currents for all points downstream of the main service panel, using the point-to-point procedure. By using this method, one can determine the available fault currents at all points throughout the distribution system with a reasonable degree of accuracy.

Methods are available that use the power of modern-day computers to make these calculations. Short-circuit current analysis of a very large or complicated distribution system should be done by computer.

REFERENCES

National Electrical Code 1990. NFPA 70 1990 Edition. Quincy, MA: National Fire Protection Association.

National Electrical Safety Code. ANSI C2-1990. New York: Institute of Electrical and Electronics Engineers.

Electrical Construction Materials Directory (Green Book). Northbrook, IL: Underwriters Laboratories.

General Information for Electrical Construction, Hazardous Location, and Electric Heating and Air Conditioning Equipment (White Book). Northbrook, IL: Underwriters Laboratories.

SPD Electrical Protection Handbook. St. Louis. MO: Bussman Company.

4

Grounding of Electrical Distribution Systems and Electrical Equipment

Gregory Bierals

DEFINITIONS

This discussion begins with the definitions for "ground" and "grounded," terms referenced in Article 100 of the National Electrical Code (NEC) (*National Electrical Code* 1990). A "ground" is defined as: "a conducting connection, whether intentional or accidental, between an electrical circuit or equipment and the earth, or to some conducting body that serves in place of the earth." "Grounded" is defined as: "connected to the earth or to some conducting body that serves in place of the earth." Thus, when the earth connection is made, the electrical circuits and equipment are considered to be grounded.

FUNCTIONS OF A GROUNDED CONNECTION

Initially an earth connection is provided for lightning protection or protection against induced transients. These transients may be due to lightning-induced impulses or may result from a utility company opening and closing a circuit switcher in a near or a distant operating station or substation. The transients may travel for miles through a distribution system. Obviously, the more connections to earth that are made the better this protection becomes. In effect, the earth is being used as a terminal to provide the protection; this is the prime reason why NEC Article 250-23(a) specifies that if a transformer is located outside a building, a grounding electrode connection shall be made to the grounded service conductor supplying the building at the transformer or elsewhere outside the building.

The second function of the earth connection is to hold the grounded conductor and the equipment grounding system at, or near, earth potential. When the earth connection is established, these two systems are referenced to a zero volt reference point. It is of the utmost importance, especially from a safety standpoint, to limit the potential difference between the grounded and grounding conductors and the earth. This potential difference will be determined by the size, composition, and length of the system grounding (grounding electrode) conductor.

During fault conditions (i.e., lightning, power faults, ground faults, etc.), the conductor that references the grounding systems to earth can carry an appreciable amount of current. It is the voltage drop associated with this fault current that establishes the voltage difference between these grounded and grounding conductors and the earth. For instance, one may consider an equipment grounding conductor (metal raceway or wire) connected to the frame of a piece of electrical equipment. At some point the ungrounded conductor supplying this equipment makes contact with the equipment frame. The ground-fault current returns through the equipment grounding system to the point where the conductor cross-bonds to the grounded conductor. Typically the cross bond would be the main bonding jumper in the service equipment (NEC 250-79d). Then the current continues through the grounded service conductor to the service transformer, completing the circuit at that point. If the impedance of this fault path were sufficiently low and the ground-fault current high enough, the circuit overcurrent protective device supplying the equipment would activate. Most of the fault current (probably about 95 percent) would flow through this path because it is the circuit with the lowest impedance.

However, some of the current can flow through the grounding electrode conductor into the earth connection (metal water pipe, ground rod, etc.). Then it can flow through the earth to the grounding electrode of the service transformer, and then through its grounding electrode conductor into the transformer. (See Figure 4.1.) This second fault path is of much higher impedance than the first, and only a limited amount (probably 5 percent) of the current will flow through this part of the circuit. It is the purpose of this earth connection to hold the equipment grounding system at or near earth potential so that the voltage drop along the equipment grounding conductor and the grounding electrode conductor will determine the voltage difference (voltage rise) between the equipment frame and the earth. This difference will be the "touch voltage potential." It is important to limit this voltage rise for safety considerations.

The term grounded conductor, defined above, is more correct to use than

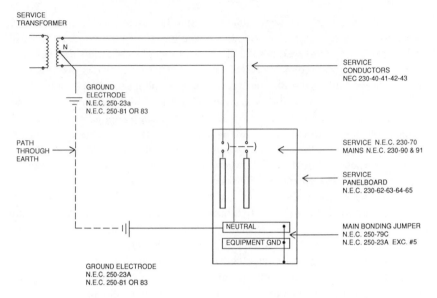

Figure 4.1. The return pathway through the grounded conductor and the earth.

"neutral conductor." All neutral conductors are grounded conductors, in that they are intentionally grounded. However, not all grounded conductors are neutrals. The neutral point of an electrical system is that point where the voltage from all other points to the neutral point will be the same, such as in the three-phase wye system or the single-phase three-wire system. (See Figure 4.2.) Here there is a neutral point, and the conductor that attaches to this point can be referred to as a neutral conductor. In a three-phase four-wire delta system and the corner-grounded delta system, the grounded conductor is not a neutral because the voltages from all other points and this grounded point are not the same. Whether the grounded wire is a neutral or not, color coding of white or natural gray must be employed in accordance with the NEC, Articles 200-6 and 200-7.

THE GROUNDING ELECTRODE SYSTEM

Purpose

The purpose of the grounding electrode system, whether it consists of a single electrode (ground rod) or multiple electrodes bonded together (ground rod, metal water pipe, concrete-encased electrode, etc.) is twofold: (1) to use the earth as a terminal to reduce the transient effects of lightning and other external power

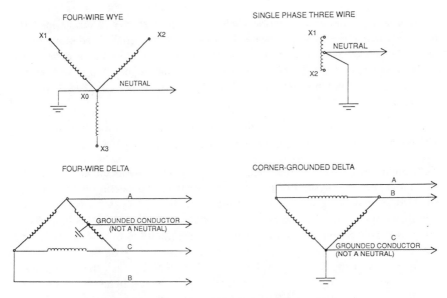

Figure 4.2. Examples of electrical system neutral points.

faults and (2) to hold the grounded and grounding system at, or near, earth potential.

Types of Grounding Electrodes

In Sections 250-81 and 250-83 of the NEC, seven different types of grounding electrode systems are defined. In Section 250-81, these systems consist of: (1) metal underground water pipes, (2) building steel, (3) concrete-encased electrodes, and (4) ground rings.

A Metal Underground Water Pipe—NEC 250-81(a)

The metal underground water pipe must be in direct contact with the earth for 10 feet or more. It must be a continuous metal pipe, free of any insulating sections such as nonmetallic pipe or couplings. This type of grounding electrode typically results in a low-resistance connection to the earth because of the extensiveness of the metal pipe in direct contact with the earth. This metal water pipe grounding electrode must be supplemented by another type of grounding electrode, of a type referenced in Sections 250-81 or 250-83, so that if the metal water pipe ever becomes inadvertently disconnected and replaced with a nonmetallic section, the grounding electrode system protection will not be lost. If the supplemental electrode is a made electrode, such as a ground rod or a buried metal plate (Article 250-83c or d), the largest conductor required to be connected to this "made electrode" is #6 copper or #4 aluminum. Installing a very large conductor would not necessarily yield a better system, but could lead to termination problems.

Effectively Grounded Building Steel—NEC 250-81(b)

The building steel may be connected effectively to earth by its direct connection to a grounding grid system or possibly by effectively connecting the structural steel to the reinforcing steel rods in a concrete footing. It is extremely important that the peripheral steel columns (the outer curtain of vertical steel columns) be effectively bonded together to ensure electrical continuity from section to section and to maintain a constant potential around the periphery of the structure. I recommend an exothermically welded conductor to accomplish this continuity, especially for Class II buildings (those that are over 75 feet in height; see NFPA 78—Lightning Protection Code). The importance of this continuity is in lightning protection. If a structure carries the surge currents of lightning, the voltage wave associated with this condition can double in magnitude at a point where the structural steel is not effectively continuous, and this voltage wave then will be reflected back into the steel. The lightning surge energy is blocked from the interior structural members by circulating loop currents through the horizontal and vertical sections of structural steel. This creates a natural blocking means to prevent the passage of this surge energy to the interior sections of the structure. For this reason it would be better to locate sensitive equipment and components, such as communications equipment and data processing or process control equipment, and sensitive instrumentation nearer to the center of the structure to avoid any association with the outer structural steel columns.

Concrete Encased Electrode—NEC 250-81(c)

This electrode sometimes is referred to as an Ufer ground. It may consist of at least 20 feet of one or more steel reinforcing bars or rods of not less than 1/2 inch diameter or not less than 20 feet of bare copper conductor not smaller than #4 AWG, encased in at least 2 inches of concrete and located near the bottom of a concrete foundation or footing. The concrete becomes a semiconducting medium because of its hygroscopic nature, and the reinforcing steel or #4 AWG or larger conductor would be electrically connected to the earth through the moisture absorbed by the concrete and the effects of capacitive coupling.

Therefore, it is very important to protect the concrete footing or foundation from damage by attaching a conductor, 2/0 AWG or larger, to an external electrode, such as a ground rod, in order to provide a low-impedance path to earth. See IEEE F77-115-9. If a steel structure were to carry severe lightning currents, and these currents were carried through the absorbed moisture in the concrete footing or foundation, the expansion of this moisture as it heated and possibly changed to steam might destroy the concrete. The external connection to the ground rod would provide an effective path for this surge energy and thus would protect the footing. It also is important to ensure adequate bonding between the structural steel and any reinforcing members in the footing. The support bolts (Jbolts) that attach to the base plate structural steel should be bonded to the steel reinforcing rods. This may be accomplished by tack-welding a section of rod to the support bolts and using the normal tie wire method of bonding reinforcing rod to reinforcing rod.

This assured continuity relieves the effects of electrolysis because of the slight rectification of the normal circulating alternating currents into direct currents (about 0.01 percent). The slight current rectification is caused by the normal circulating currents flowing through metallic oxides, which form around the concrete-encased reinforcing rods and into the concrete. The direct current contributes to the electrolysis condition, and may lead to accelerated corrosion of the rebar; and as the steel corrodes and expands in volume, excessive internal pressure is exerted on the concrete, which can lead to its destruction. (See IEEE STD. 80—Concrete Encased Electrodes.)

The Grounding Ring—NEC 250-81(d)

This grounding electrode consists of a #2 AWG or larger conductor that completely encircles a building at a depth below the earth's surface of 2½ feet. This wire typically is bonded to driven ground rods to effectively lower its resistance. The primary function of the ground ring is to equalize the potential gradient around a structure during lightning conditions. According to the Lightning Protection Code (NFPA 78) the minimum size of the main bonding conductor, which connects the lightning air terminals around the top of a building, is required to be at least 57,400 circular mils for a Class I building (a building up to 75 feet in height) (#2 AWG is 66,360 circular mils), or for a Class II building (over 75 feet in height) 115,000 circular mils (2/0 AWG is 133,100 circular mils). This wire size also is required for the bonding conductors used to provide a two-way path to ground from the air terminals and for the main bonding conductors that are installed at about 60- to 100-foot horizontal intervals in order to equalize potential differences between lightning down conductors, as well as for the grounding ring conductor installed in the earth around the structure. Of course, local soil conditions may cause these conductors to be increased in size owing to the effects of corrosion. (See NFPA 78 for information regarding the stranding of these conductors in Chapter 3, Tables 3-4 and 3-5).

Made and Other Electrodes—NEC 250-83

According to Article 250-83, if none of the electrodes referenced in Article 250-81 is available, one or more of the electrodes listed in Article 250-83 may be used. A "made electrode" is one that is installed solely for grounding purposes such as a ground rod, buried metal pipe, and so on. These electrodes include: (1) local metal underground systems, and (2) rod and pipe electrodes.

Metal Underground Gas Piping System—NEC 250-83(a)

The NEC prohibits the use of this type of piping system as a grounding electrode. This gas pipe probably would be a good, low-resistance grounding electrode if a significant section of the metallic surface were in direct contact with the earth. However, because of the effects of corrosion, this pipe typically is coated with coal tar or other bitumastic coating for protection, and very little direct earth contact is established. The local gas supplier may be employing an impressed-current cathodic protection system for additional protection against the effects of galvanic currents from nearby dissimilar metals. This form of cathodic protection is very common where extensive metal piping or metallic vessels are installed underground, and the cost of operating this system is greatly reduced if the metallic surface area is well coated. Usually the amount of current circulating through the earth is based on 2 to 20 mA for each square foot of bare steel exposed to the earth when an impressed-current cathodic protection system is used. Also, a gas pipe may utilize insulating-type couplings or joints that would affect its resistance to earth. For these reasons, the NEC in Section 250-83(a) prohibits the use of metal underground gas pipe as a grounding electrode.

NFPA 54-1984, the National Fuel Gas Code, states that the gas piping shall not be used as a grounding electrode. It is recommended that all other metallic piping and metal air ducts within the premises be bonded together in order to maintain a constant potential among the various systems throughout a building or structure. This also is indicated in NEC Article 250-80 (FPN). If one system is subjected to a voltage excursion due to making contact with an energized conductor or a lightning discharge, the bonding will serve to hold all metallic systems to approximately the same potential. However, some small potential differences would exist due to the voltage drop associated with the various bonding conductors and their relative lengths.

Other Local Metal Underground Systems or Structures, Such as Piping Systems and Underground Tanks—NEC 250-83(b)

These metallic structures typically would serve as an excellent grounding electrode because of their extensive area; but these types of piping systems usually are covered with a nonconductive coating, and no direct earth contact would be made. Therefore, the resistance to earth of this grounding system would be too high to be effective.

Rod and Pipe Electrodes—NEC 250-83(c)

The minimum depth of a driven pipe or rod is specified as 8 feet. Depth plays an important role in the resistance to earth of the driven electrode because one of the most important aspects of earth resistivity is the effect of the moisture content by weight, and as the electrode depth increases, this moisture content usually increases. Sometimes it is impossible to drive the pipe or rod to an 8-foot depth because rock bottom is encountered at higher elevations. If this is the case, the pipe or rod may be driven at an oblique angle not to exceed 45° from the vertical, or the electrode may be buried horizontally in a trench that is at least $2\frac{1}{2}$ feet deep. The upper end of the electrode shall be flush with or below ground level so that the grounding electrode conductor attachment is inherently protected from physical damage. If the upper end of the pipe or rod and the grounding conductor is exposed, then a protective covering generally is required (NEC Article 250-117b). The diameter of the pipe or rod has little to do with its overall resistance.

The NEC requires the diameter of a galvanized pipe to be at least 3/4 inch. For a steel rod, the minimum diameter is 5/8 inch, and for a nonferrous rod the minimum diameter is 1/2 inch. It is important that the rod's tensile strength be sufficient to resist bending of the rod. This is the major reason why the NEC specifies a minimum diameter of 1/2 inch nonferrous and 5/8 inch ferrous rod. However, driving rods of a larger diameter than necessary is wasteful. A rod one inch in diameter will have about $9\frac{1}{2}$ percent less resistance to earth compared to a rod with a diameter of 1/2 inch but will increase cost and weight about 400 percent. The three components of the total grounding electrode resistance are: (1) the resistance of the mass of the electrode, (2) the metal-to-soil interface area, and (3) the earth's resistivity (discussed in the following paragraphs).

The Resistance of the Mass of the Electrode

For a typical 10-foot-long by 5/8-inch-diameter steel rod, the electrical resistance of the metallic mass certainly would be negligible—probably a few thousandths of an ohm.

The Metal-to-Soil Interface Area

This component of the grounding electrode resistance also is negligible because of the limited contact area between the metallic surface and the earth. However, if the earth temperature is quite low, this part of the electrode resistance cannot be ignored. The earth resistivity is directly affected by the earth temperature. As the temperature drops, the earth resistivity increases. This is another reason why depth is important. As moisture forms around the surface of the pipe or rod, a veneer of ice also will form under freezing conditions; and when this situation occurs, the metal-to-soil interface will be lost. For example, the earth resistivity may increase 300 percent if the earth temperature drops from 50°F to 32°F. Unfortunately, it is not always possible to drive pipes and rods to great depths.

It may be necessary at times to drive electrodes in parallel at shallower depths, an arrangement that may be acceptable if the electrodes are spaced properly. The paralleling efficiency is dependent on proper spacing between the rods. As a rule of thumb, the space between parallel rods is determined by a value of twice their driven depth; for example, two rods that are 10 feet deep should be spaced 20 feet apart (see Figure 4.1). Two rods in parallel, effectively spaced, may have approximately 60 percent of the resistance of a single rod, three rods approximately 40 percent, and four rods approximately 33 percent. An alternative may be to install buried copper wires at very shallow depths. Of course, the length of these buried conductors is important in establishing adequate surface contact with the soil. For example, NFPA 78, the Lightning Protection Code, would require each down conductor of a lightning protection system to extend at least 10 feet vertically into the earth. If bedrock is near the surface, the down conductor is to extend at least 12 feet in a horizontal plane away from a building at a depth of one to two feet in clay soil, or it will be 24 feet long and 2 feet deep in sandy or gravelly soil. If the soil is less than one foot deep, then a loop or ring conductor is to be installed, and the lightning down conductors are to be connected to the loop conductor. The size of the loop or ring cable is to be the equivalent of a main size lightning conductor, which would be #2 AWG copper for a Class I building or 2/0 copper for a Class II Building. According to ANSI C2-1990, the National Electrical Safety Code, the minimum diameter of a buried conductor is 0.162 inch (#6 AWG has a diameter of 0.184 inch) and the length of this buried grounding conductor is 100 feet. The minimum depth is 2 feet, but shallower depths are permitted where this depth is impossible. This buried conductor is referred to as a "counterpoise," and sometimes these wires are run in multiples with their ends and midpoints joined together.

For ground conditions similar to those expected in a permafrost setting, a horizontal wire is preferable to a vertical electrode. Possibly a mixture of soil, salt, and water may be poured around the wire to increase the effectiveness of the grounding installation in permafrost.

Another alternative is either to use concrete-encased electrodes or to surround the buried grounding conductors with bentonite, a form of natural clay. Both concrete and bentonite have resistivities lower than that of the earth under ideal conditions, and these substances surrounding the metallic mass of the electrode tend to lower its resistance by increasing the overall diameter of the grounding electrode.

An important element that should not be overlooked is the effect of corrosion. Four conditions are necessary for corrosion:

1. Two dissimilar metals are required.
2. The metals must be electrically connected to each other.
3. The metals must be surrounded by an electrolyte.
4. Oxygen is necessary.

If one of the four requirements is not present, corrosion is eliminated. The rate of corrosion decreases with time owing to the effects of polarization. When two dissimilar metals are connected through an electrolyte (e.g., steel as an anode and copper as a cathode, with the earth separating them as the electrolyte), a galvanic or corrosion current flows between them (in this case from the steel to the copper). The result is a voltage change on both the anode and the cathode surfaces. Such potential changes are called polarization. The polarization effect causes the anodic potential to become more cathodic and the cathodic potential to be more anodic. This phenomenon results in a decrease of galvanic cell voltage and of the corrosion current, which can drop to less than 10 percent in about 80 to 100 hours. If it were not for polarization, the deterioration of dissimilar metals in the earth would be extremely rapid.

If ground rods are to be protected from corrosion, they should have a copper sheathing metallurgically bonded to the steel core. This atomic interconnection of the metals is important for several reasons. If the rod were to strike a rock as it was being driven and the driving force caused it to bend, the copper sheathing would not split and allow moisture to surround the steel core. The oxygen element of this moisture would accelerate corrosion of the rod. According to UL Spec. No. 467, the minimum copper thickness is 0.25 MM. This copper protection is more than sufficient for an average service life of 30 years.

Another effect on the corrosion of the rod is due to the special phenomenon of fault currents being discharged through the rod into the earth. During initial discharges the current is uniformly distributed over the entire surface of the rod. If the duration of the discharge current is more than a few milliseconds, the earth surrounding the rod becomes so dry that an arc is formed at one point on the rod. The soil temperature will increase at this point, and the soil will dry out even more; and as the soil resistivity increases, the fault current will seek another location along the rod, and another arc will develop. This phenomenon will continue until a circuit protective device operates to clear the fault; and at the same time the ground rod will experience temperatures well above the boiling point of zinc, but normally not above the boiling point of copper. Under these fault conditions, a galvanized-only rod will lose its zinc protectorant and thus its corrosion resistance.

In addition, all grounding electrodes carry some current under normal conditions. Even a very small current of 50 to 200 mA, where the soil has a relatively high resistivity, can result in a rod surface temperature greater than the boiling point of zinc. Once again, the loss of zinc protection would accelerate the corrosion of the rod. In summary, a copper-clad steel rod may last 8 to 10 times as long as a galvanized steel rod or pipe.

Stray DC currents also affect corrosion. All buried metals are subject to a buildup of metallic oxides on their surface. As AC leakage currents are carried through these metal oxides, some AC current is rectified into DC. These rectified currents cause more rapid corrosion of zinc, iron, and steel than they do with copper.

The Resistivity of the Soil or the Medium Surrounding the Electrode

This component of the grounding electrode resistance is by far the major portion of the total resistance. The earth's resistivity is affected by:

- Temperature
- Salt
- Moisture content by weight

First, as the temperature of the earth increases, the resistivity decreases. For example, for a sandy loam soil with a moisture content of approximately 15 percent, the resistivity would be about 7200 ohm-centimeters at 68°F, as compared to 9900 ohm-centimeters at 50°F.

Second, the salt content of the earth also affects the earth resistivity, and adding salt to the soil can make up for such seasonal conditions as earth temperature changes or moisture variations. The most common salting materials are:

- Sodium chloride
- Magnesium sulfate
- Calcium chloride
- Copper sulfate

Sodium chloride and magnesium sulfate are the most commonly used of these materials. Salt may be added to the soil by digging a trench around a ground rod, approximately 18 inches away from the rod and one foot deep. The trench would be filled with 40 to 70 pounds of salt, which would migrate into the soil during periods of rainfall. Salt in this quantity usually will last for several years before replacement is necessary. Also, a hollow-core rod is available. This core is filled with soluble chemicals. These chemicals react with normal moisture in the air, dissolve, and migrate into the soil. Once again, the chemicals may be replaced from time to time. The effects of salt in decreasing earth resistivity level off when the content reaches about 6 percent. This amount of salt is not overly corrosive, but extreme amounts may accelerate the corrosion process.

Third, the moisture content by weight directly affects the earth resistivity. Considering the same sandy loam soil with a fairly constant temperature, the resistivity would be approximately 150,000 ohm-centimeters at a $2\frac{1}{2}$ percent moisture content, as compared to 6,300 ohm-centimeters at a 20 percent moisture content. The effects of moisture on soil resistivity level off above 20 percent in that a change from 20 percent to 30 percent results in a difference of only 2,100 ohm-centimeters (6,300 ohm-centimeters at 20 percent vs. 4,200 ohm-centimeters at 30 percent). As discussed earlier, if the moisture in the earth freezes, the earth resistivity increases dramatically; so driven electrodes must be deep enough, at least below the frostline, to avoid this condition.

Plate Electrodes—NEC 250-83(d)

Each plate electrode is to expose at least 2 square feet of surface to the soil. These plates may be employed where bedrock is at high elevations and driven rods would be impractical. They must have a minimum thickness of 1/4 inch if they are galvanized steel or .06 inch if they are of nonferrous material. Once again, the reader should keep in mind that the steel plate typically will corrode faster than the nonferrous plate for the same reason that a steel ground rod or pipe will corrode faster than one of nonferrous material, in accordance with the previous discussion. The minimum burial depth of this plate electrode is 5 feet. These buried plates typically are installed on end rather than flat to reduce the amount of excavation required.

Magnitude of Resistance to Ground

Article 250-84 indicates that the maximum resistance to a ground of a "made electrode" is not to exceed 25 ohms. A water pipe is not a made electrode because it is not installed solely for use as a grounding electrode. It is important to note that a 25-ohm value is not something one should strive to achieve. It is always desirable to have a much lower resistance to ground. The reason for this is illustrated in Figure 4.3. If there were to be a current injection of 1,000 amperes into a 25-ohm ground rod, then 25,000 volts would be dropped around the rod, and about half of this voltage would be present in about a 1/2-foot radial area around the rod. The system certainly would be safer if the resistance were 5 ohms or less. As noted earlier, the two functions of the grounding electrode system are to provide protection and to hold the grounded conductor and the equipment system at, or near, earth potential.

Distance from rod surface - Approximate Percentage of Resistance

0.1'	-	25%
0.2'	-	38%
0.5'	-	52%
1.0'	-	68%
5.0'	-	86%
10.0'	-	94%
25.0'	-	100%

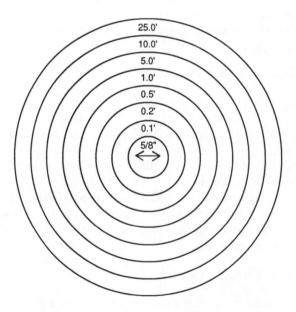

Figure 4.3. The sphere of influence of a ground rod.

SELECTING THE PROPER WIRE SIZE FOR GROUNDING AND GROUNDING ELECTRODE CONDUCTORS

The primary purpose of grounding systems is to provide a low-impedance electrical path to ground. Therefore, it is necessary to correctly select the wire sizes of the various conductors that compose the grounding system. If an installed grounding or grounding electrode wire is too small, the impedance path to ground will be too large, and safety will be compromised. Unfortunately, there is confusion in the industry about how to select the correct wire size for these components. The following discussion is presented to eliminate this confusion.

The National Electrical Code contains tables relating to the size of grounding electrode conductors for AC systems (Table 250-94) and equipment grounding and bonding conductors (Table 250-95). These two tables were based on data provided in an IEEE committee report, "A Guide to Safety in AC Substation Grounding." This report discusses the validity of the grounding conductor sizes specified in Tables 250-94 and 250-95, based upon a typical length of 100 feet and the grounding conductor voltage drop associated with this length. A criterion for the maximum voltage drop acceptable for any length of wire can be developed by using the DC resistance values specified in the NEC, Chapter 9, Table 8, entitled "Conductor Properties." Table 8 specifies DC resistance values for coated (tinned) copper, uncoated copper, and aluminum wire based on a 75°C ambient temperature and a conductor length of 1,000 feet.

An explanation of the use of coated (tinned) copper is in order. The tinning of copper was necessary to reduce the effects of corrosion that normally occurs when rubber-based insulation is used directly over copper. There is an intrinsic chemical reaction between rubber and copper that produces the corrosion. This reaction is not present when modern thermoplastic insulation is used.

Tinning may be necessary where bare conductors are exposed to corrosive atmospheres, or where the bare copper conductors are directly buried in chemically active soils. The DC resistance values are slightly higher for coated than for uncoated copper.

There is another consideration in the selection of grounding conductors with regard to the wire sizes specified in Tables 250-94 and 250-95—the thermal stress or short-time heating rating of these conductors based on a 5-second current-carrying capacity. The thermal stress is represented mathematically by the following formula:

$$\text{Thermal stress} = I^2 \times T$$

where I equals the rms current in amperes, and T equals the time in seconds. The 5-second withstand rating of copper conductors is considered to be the equivalent heating of one ampere current flow for every 42.25 circular mils of conductor cross-sectional area (Fink 1987).

Therefore, a 5-second withstand current rating for each copper wire size can be calculated by dividing the cross-sectional area in circular mils by 42.25:

$$\text{5-second rating} = \frac{\text{circular mil area}}{42.25}$$

Example: Five-Second Withstand Rating Calculations

Applying the above equation to the various copper wire size data provides the following 5-second withstand current ratings:

$$\#14 \text{ wire: } \frac{4,110 \text{ cmil}}{42.25} = 97 \text{ A}$$

$$\#12 \text{ wire: } \frac{6,530 \text{ cmil}}{42.25} = 155 \text{ A}$$

$$\#10 \text{ wire: } \frac{10,380 \text{ cmil}}{42.25} = 246 \text{ A}$$

$$\#8 \text{ wire: } \frac{16,510 \text{ cmil}}{42.25} = 391 \text{ A}$$

$$\#6 \text{ wire: } \frac{26,240 \text{ cmil}}{42.25} = 621 \text{ A}$$

$$\#4 \text{ wire: } \frac{41,740 \text{ cmil}}{42.25} = 988 \text{ A}$$

$$\#3 \text{ wire: } \frac{52,620 \text{ cmil}}{42.25} = 1,245 \text{ A}$$

$$\#2 \text{ wire: } \frac{66,360 \text{ cmil}}{42.25} = 1,571 \text{ A}$$

$$\#1 \text{ wire: } \frac{83,690 \text{ cmil}}{42.25} = 1,981 \text{ A}$$

$$\#1/0 \text{ wire: } \frac{105,600 \text{ cmil}}{42.25} = 2,499 \text{ A}$$

By multiplying these 5-second withstand current ratings by the DC resistance values of 100-foot lengths of wire, one can determine the maximum touch potential exposure during a fault condition. (See the example calculations that follow.)

Touch Voltage

The DC resistance of #8 uncoated stranded copper wire equals 0.778 ohm per 1,000-foot length at 75°C (Table 8, Chapter 9, NEC). Thus, a 100-foot length of wire would have a resistance of 0.0778 ohm. Multiplying this resistance by the 5-second withstand current of 391 A (calculated above) yields a voltage drop of 30.42 V for the 100-foot length of wire at the withstand current (0.778 ohm × 391 A = 30.42 V). A corresponding calculation for any of the other wire sizes will yield a similar value for the voltage drop, that is, approximately 30.5 V. This is generally assumed to be a safe touch potential level. The wires sizes listed in Tables 250-94 and 250-95 are based upon limiting the voltage drop and the touch potential during fault conditions; also they are based upon a maximum conductor length of 100 feet. Grounding electrode conductor lengths exceeding 100 feet have to have their wire size adjusted accordingly. (See the following example for a calculation procedure.)

Example: Adjusting for Conductors Longer than 100 Feet

Given a situation with a 150-foot-long grounding electrode conductor and #2 AWG copper service conductors, one must find the proper wire size for the grounding electrode conductors.

Solution. Table 250-94 lists a copper wire size of #8 AWG for the grounding electrode conductor. However, this table assumes the conductor run to be no more than 100 feet. In order not to exceed a voltage drop of 30.5 V, the resistance of the wire size chosen for the grounding electrode conductor must not exceed the 0.0778 ohm calculated previously, and a wire size larger than a #8 is required.

To determine the proper wire size, the maximum allowed resistance of 0.0778 ohm is divided by a factor of 0.15 (0.0778/0.15), to give 0.52 ohm/1,000 feet. This limits the wire size to one having a maximum resistance of 0.52 ohm per 1,000-foot length. (The factor 0.15 is used because the length of the conductor is 150 feet, or 15% of 1,000 feet.) The resistance values of Table 8 are listed for 1,000-foot lengths of wire. A lookup in Table 8 shows a minimum wire size of #6 uncoated copper wire (0.491 ohm per 1,000 feet) for the grounding electrode conductor.

SIZING EQUIPMENT GROUNDING CONDUCTORS

Equipment grounding conductors should be sized by the same procedure in order to minimize the voltage drop during fault conditions. This method for selecting the proper wire size for grounding electrode and equipment grounding conductors may be used to minimize the potential voltage rise on both grounded conductors and the equipment grounding system (i.e., equipment frames, conduits, metal boxes, etc.). The procedure complies with the provisions of NEC Article 250-51, entitled "Effective Grounding Path." This article lists three distinct requirements for an effective grounding path for circuits, equipment, and conductor enclosures; it must:

1. Be permanent and continuous.
2. Have the capacity to conduct safely any fault current likely to be imposed.
3. Have sufficiently low impedance to limit the voltage to ground and to facilitate the operation of the circuit protective devices in the circuit.

In addition, the earth shall not be used as the sole equipment grounding conductor.

Conductor Selection Based upon Short-Circuit Withstand Ratings

Another important consideration for correctly sizing equipment grounding conductors is their short-circuit withstand capability. Withstand ratings are defined in ICEA Publication P-32-382. A solid knowledge of the parameters of a circuit is required to select the proper wire size for an equipment grounding conductor. Among these parameters are:

1. The operating characteristics of the circuit overcurrent protective device; the total clearing time; current limiting characteristics, if any; and others.
2. The available short-circuit fault currents at all points in the circuit. These values must be determined by a careful short-circuit study of the electrical system.
3. Other circuit characteristics.

These requirements are mandated by NEC Article 110-10, "Circuit Impedance and Other Characteristics." This article requires that the circuit protective devices be chosen so that a fault may be cleared before extensive damage occurs to the components of the circuit. One of the main circuit components, which is often overlooked, is the wire or cable, which must be properly sized to withstand any available fault current. This is particularly important for equipment grounding conductors if they are to limit the potential rise on any equipment during a fault condition. The following example illustrates the procedure for selecting the proper equipment ground wire size.

Example: Selecting an Equipment Grounding Conductor

The motor of Figure 4.4 is fed through the combination starter from a wire as shown. The available fault current at the wire way has been determined to be

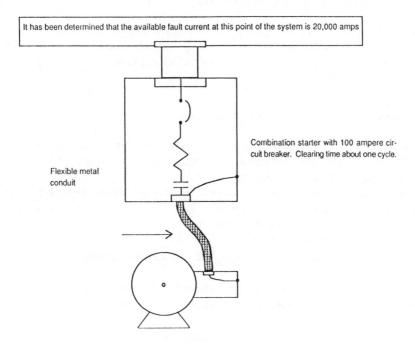

Figure 4.4. Example of equipment grounding conductor.

20,000 sca. The 100 A circuit breaker (CB) in the starter has a clear time of about one cycle or 1/60 second.

The equipment ground conductor suggested by Table 250-95 is a #8 copper wire. This selection is based on a CB size of 100 A.

However, upon consulting the ICEA publication (P32-382), one finds that the withstand rating of a #8 copper wire for one cycle is approximately 6,900 A. This is considerably less than the 20,000 sca available. Therefore, in order to satisfy the NEC, Section 110-10, a larger wire size for the equipment grounding conductor is required. As a minimum, it must have a withstand rating of 20,000 sca for one cycle.

Additional consultation of Publication 32-382 yields the 20,000 sca for one cycle based upon an operating temperature of 75°C and a maximum short-circuit temperature of 150°C. In this case, a #3 AWG copper wire would be required for the equipment grounding conductor.

If the overcurrent protective device is a current-limiting fuse, the wire size for the equipment grounding conductor usually can be chosen directly from Table 250-95. (See the discussion on current-limiting fuses that follows.)

CONDUCTOR WITHSTAND RATINGS

The concept of conductor thermal stress was introduced previously. In that discussion, 5-second withstand current ratings were calculated for a variety of conductor wire sizes. In a similar manner, withstand ratings can be calculated for any time period.

Example: Withstand Current Calculation for #10 AWG

A #10 AWG copper wire will be used as an example for the calculation of a withstand rating for any time period. The method can be expanded by the reader to any copper wire size of interest.

It was shown earlier that the #10 copper wire had a 5-second withstand rating of 246 A. Substituting the values of $I = 246$ A and $T = 5$ seconds into the thermal stress equation yields the following:

$$I^2 \times T = 246^2 \times 5 = 302,580 \text{ ampere}^2\text{-seconds}$$

This ampere^2-seconds value now can be used to calculate the withstand current rating for other time periods. Solving the above equation for I yields:

$$I = \sqrt{(302,580/T)}$$

For a time period, T, of one second, the expression becomes:

$$I = \sqrt{(302,580/1)} = 550 \text{ A}$$

Thus, the one-second withstand current rating for a #10 AWG copper wire is 550 A.

For a time period of one cycle or 1/60 second:

$$I = \sqrt{(302,580/0.0167)} = 4,257 \text{ A}$$

The one cycle withstand current rating for a #10 AWG copper wire is 4,257 A.

For a time period of one-half cycle:

$$I = \sqrt{(302,580/0.0083)} = 6,150 \text{ A}$$

The one-half cycle withstand current rating for a #10 AWG copper wire is 6,150 A.

These examples demonstrate that the short-time, short-circuit withstand rating of a conductor is easily calculated. Thus, it is easy to check to make sure the conductors and other circuit components can safely withstand the fault currents predicted by the short-circuit study of the distribution system.

Note. Remember that all circuit components are required to have appropriate withstand properties. This applies to the equipment grounding conductor system and associated conducting paths and all other ungrounded circuit components. (See NEC, Section 110-10.)

LET-THRU CURRENT AND OVERCURRENT PROTECTION DEVICES

Current-Limiting Fuse

Another method of ensuring compliance with these requirements is to check the fault–let-thru current of the circuit device and compare the I^2T (thermal stress) rating with that of the circuit components. For instance, assume that the #10 AWG conductor in the above example was protected by a current-limiting fuse. Is the conductor adequately protected by the fuse for a fault condition? An examination of the fuse manufacturer's data sheet will provide the value of the fault–let-thru current. A comparison of the two I^2R ratings will provide the answer to the question.

Example: Current-Limiting Overcurrent Protection Device

Suppose the current-limiting fuse was listed as having a fault–let-thru current of 4,000 A, based upon a 100,000 A available fault current and a total clearing time of one-quarter cycle. The I^2T let-thru energy of the fuse is easily calculated as:

$$\text{(Fuse) } I^2T = 4{,}000 \text{ A} \times 4{,}000 \text{ A} \times (1/240) \text{ seconds}$$

$$= 66{,}667 \text{ ampere}^2\text{-seconds}$$

Similarly, the I^2T rating of the #10 AWG copper for 5 seconds is calculated as:

$$\text{(\#10 AWG) } I^2T = 246 \text{ A} \times 246 \text{ A} \times 5 \text{ seconds}$$

$$= 302{,}580 \text{ ampere}^2\text{-seconds}$$

Thus, the #10 wire is clearly protected by this fuse. This is true for the conductor whether it is the equipment grounding conductor or the ungrounded conductor of the circuit.

Example: Non-Current-Limiting Protection Device

Consider a circuit with an available symmetrical fault current of 100,00 A with a non-current-limiting overcurrent device with a one cycle opening time. What is the I^2T let-thru energy?

The I^2T let-thru energy is:

$$I^2T = 100,000 \text{ A} \times 100,000 \text{ A} \times .0167 \text{ seconds}$$

$$= 167,000,000 \text{ ampere}^2\text{-seconds}$$

Clearly, this let-thru energy is well above the short-time withstand rating of the #10 copper wire.

When the overcurrent protection device is a current-limiting fuse, such a Class RK1, T, J, and similar types, the grounding conductor wire sizes may be selected directly from Tables 250-94 and 250-95. This is so because the fast response of the current-limiting fuse to short circuits and ground faults reduces the time that the fault will endure and the total energy passed through to the conductor.

In summary, not only is it necessary to protect conductors and components in agreement with their respective ampacities, but they also need to be protected against short-circuit and ground-fault conditions as well.

PHYSICAL PROTECTION OF GROUNDING ELECTRODE CONDUCTORS

The question of how to protect the grounding electrode conductor from physical damage is of particular interest. Section 250-92(b) states that where metal race-ways are provided for protection of the grounding electrode conductors, they shall be made electrically continuous from the point of attachment to cabinets or equipment to the grounding electrode. If a steel conduit sleeve is used for physical protection, some means must be provided on each end of the sleeve to make it electrically continuous. This can be accomplished by installing a bonding jumper on each end and terminating the bonding jumper to the equipment and to the grounding electrode. The reason why this bonding method is important is that during heavy fault conditions the steel conduit produces a choke effect. This results in a significant increase in the impedance of the grounding path. IEEE Paper 54-244 discusses an unusual division of current that occurs during these fault conditions. The bonding of the steel raceway as described here does not eliminate this choking effect but does alleviate it by reducing the grounding conductor impedance. Therefore, it probably is advisable to provide a nonmetallic covering of physical protection wherever possible.

EFFECTIVE GROUNDING CONSIDERATIONS

The importance of establishing effective grounding paths for electrical distribution systems has been stressed. The question of whether a piece of equipment or an electrical circuit is effectively grounded is certainly one of extreme importance. However, the difficulty in determining the viability of grounding paths in a real-world environment may lead to improper and unsafe conclusions. The multiplicity of conducting metallic bodies and structure that parallel equipment grounding paths may make it difficult to determine the impedance of a particular grounding circuit. Even with appropriate test equipment this still can be a major problem. The importance of making proper and detailed calculations about the grounding paths before construction begins cannot be overstressed.

What about the thousands of circuits that were never provided with a grounding conductor? What about circuits such as the old knob and tube systems or the original nonmetallic cable installations that had no equipment grounding conductor? Also, many earlier installations relied upon connections to existing structural steel as part of the grounding circuit, a condition expressly prohibited by the NEC in Section 250-58(a) because of the increased impedance of this conducting path associated with its physical separation from the conductors supplying the connected equipment. OSHA, 29 CFR, Part 1910, Subpart S, does recognize an exception for equipment installed prior to April 16, 1981 where this condition existed, but certainly not for later installations.

Ground Fault Circuit Interrupters (GFCIs)

In some cases, the NEC recognizes the use of ground fault circuit interrupters on circuits that have no grounding conductor. Section 210-7(d) indicates that a receptacle-type GFCI may be installed to provide an added degree of personnel protection. This type of device does not rely on an equipment ground to operate.

The principles of Kirchoff's current law explain the operation of the GFCI. As long as the current supplied to the load by the ungrounded conductor is equal in magnitude to the current returning through the grounded conductor the circuit performs normally. There is no leakage current problem as long as these currents are equal and opposite. A differential transformer surrounds both of these conductors and monitors the currents in these conductors. If a current difference of 4 to 6 mA for a Class A GFCI develops, a magnetic flux is set up in the transformer core causing a solid-state trip circuit to activate in about 1/40 second. One receptacle-type GFCI may be arranged to provide protection for multiple down-stream receptacle outlets on the same branch circuit. Also, NEC Section 210-7(d) recognizes the possibility of replacing these downstream two-prong receptacles with three-prong grounding type outlets in order to accommodate three-prong male plugs even though no equipment grounding conductor has been provided. This presents no problem because the GFCI needs no equipment grounding conductor for its operation. This will allow the use of three-prong plugs with no need to trim the grounding prong, which unfortunately sometimes is necessary when two-prong receptacles are available.

The importance of effective grounding, from a safety standpoint, is the prevention of serious injury or death by limiting the amount of current available to pass through the body during a fault condition. The current passing through the body during an electric shock is determined by the voltage rise on equipment and the impedance of the body. A very small current flow, on the order of 80 mA, can cause ventricular fibrillation, which very likely will result in death. Tests have been performed to determine the maximum shock duration time at 80 mA for a typical 150-pound human being with a body impedance of 1500 ohms. The results of these tests are as follows:

Circuit Voltage	Duration (Seconds)
120	4.2
240	1.05
277	0.8
480	0.26

A 20 A molded case circuit breaker may require a current equivalent to six times its rating (120 A) to clear in 4.2 seconds. Using Ohm's law, on a typical 120 V circuit the impedance of the grounding circuit would be 1 ohm:

$$R = \frac{V}{I} = \frac{120 \text{ V}}{120 \text{ A}} = 1 \text{ ohm}$$

In this example, this circuit impedance may be considered to be safe because the time of exposure will be limited to 4.2 seconds at 80 mA. Again Ohm's law is used:

$$1,500 \text{ ohms} = \frac{0.08 \text{ A}}{120 \text{ V}}$$

Any increase in circuit impedance will extend the time of exposure and cause ventricular fibrillation to occur before the circuit breaker operates to clear the fault. Safety factors should be considered carefully because of the varying clear-ing times and operating characteristics of overcurrent protective devices. Also, metallic raceways may constitute the grounding path, and the impedance of these systems may be affected by corrosion, loose connections, or other physical damage. In contrast, supplementing these metallic raceways with internal equip-ment grounding conductors may prove beneficial in reducing impedance and the fault clearing time of the circuit overcurrent protective device.

See additional discussion on the physiological effects of electricity in Chapter 2.

REFERENCES

Fink, D. G. 1987. *Standard Handbook for Electrical Engineering, 12th Edition*. New York: McGraw-Hill.

IEEE STD. 80. A Guide to Safety in AC Substation Grounding. New York: Institute of Electrical and Electronics Engineers.

The Lightning Protection Code. NFPA 78 1990 Edition. Quincy, MA: National Fire Protection Association.

The National Electrical Code. NFPA 70 1990 Edition. Quincy, MA: National Fire Protection Association.

The National Fuel Gas Code. NFPA 54 1984 Edition. Quincy, MA: National Fire Protection Association.

The National Electrical Safety Code. ANSI C2-1990. New York: Institute of Electrical and Elec-tronics Engineers.

5

OSHA Standards and Requirements and the National Electrical Code

Gregory Bierals

INTRODUCTION AND HISTORICAL PERSPECTIVE

The Occupational Safety and Health Administration (OSHA) has determined that electrical hazards in the workplace pose a significant risk of injury or death to employees. As a means of providing protection against these electrical hazards, OSHA has established a document, *Design Safety Standards for Electrical Systems,* herein referenced as 29CFR Part 1910, Subpart S. This standard was revised and became effective on April 16, 1981. The revision was considered necessary to simplify as well as to update the former standard. The regulations draw heavily upon the experience of the National Electrical Code and specifically on the relevant requirements of the NEC.

In the previous OSHA electrical standard, the NEC was incorporated by reference. Now, employers can refer directly to the text of the OSHA regulations to determine their obligations. The OSHA electrical standard is designed to be used in correlation with the 1978 edition of the NEC. There are only a few significant differences between these two safety standards. The original OSHA electrical standard incorporated virtually the entire 1971 National Electrical Code. The revision of this standard was not intended to lessen its contribution to safety in the workplace. However, the NEC is revised every 3 years, and any specific edition that OSHA might adopt soon would become obsolete. A similar 3-year update of OSHA regulations was determined to be impractical. As a remedy to this problem, it was decided that OSHA's electrical standards should be designed to accommodate changes in technology without the need for constant revision. In addition, it was decided to write these safety requirements in a performance-oriented fashion. Another problem of adopting the entire NEC into the OSHA standard is that the NEC contains many requirements not directly related to employee safety. The National Fire Protection Association (NFPA), which is sponsor of the National Electrical Code, had several meetings with OSHA officials. It was decided that the NEC Correlating Committee should examine the possibility of preparing an electrical safety standard to be used as a basis for electrical safety in the workplace. Eventually NFPA 70E, *Electrical Safety Requirements for Employee Workplaces,* was published. This safety publication was designed around the following criteria:

Part I—Installation safety requirements
Part II—Safety-related work requirements
Part III—Safety-related maintenance requirements
Part IV—Safety requirements for special equipment

Once again, the 1978 National Electrical Code was used as the source document for NFPA 70E, but significant material referenced in the NEC was deleted because it did not directly involve employee safety. NEC Chapters 3 and 4 were deleted, as these sections deal with wiring methods and materials. These chapters contain ampacity tables and various types of installation standards regarding types of wiring methods, lighting fixtures, appliances, space heating equipment, motors, generators, and transformers. Certain specific concepts of these sections were kept; for example, the requirements that appropriate conductors be insulated and that the insulation be approved for its intended purpose, use, function, environment, and application were retained. NFPA 70E does contain certain provisions that are contained in Chapter 6 of the NEC, regarding electrically driven or controlled irrigation machines and electrolytic cells typically used in electroplating processes. These installations have various electrical hazards and so are included in 70E.

Chapter 7 generally deals with special conditions, and most of it was deleted. Those parts dealt with legally required or optional standby power systems. Employee safety typically is not directly affected by these types of systems. Several sections, however, are included in NFPA 70E. These parts concern special signaling and alarm circuits, such as fire protective signaling systems, and do affect workplace safety. Chapter 9 also was deleted. It presents technical information and example computations related to specific electrical systems, which generally are used for instruction purposes.

In summary, OSHA officials reviewed NFPA 70E, Part I and determined that it could be used as a source document for OSHA's design safety standard for electrical utilization systems and equipment. It eventually became Subpart S of 29CFR Part 1910.

The OSHA sections included in this subpart are:

1910.301—Introduction
1910.302—Electrical utilization systems
1910.303—General requirements
1910.304—Wiring design and protection
1910.305—Wiring methods, components, and equipment for general use
1910.306—Specific-purpose equipment and installations
1910.307—Hazardous (classified) locations
1910.308—Special systems
1910.399—Definitions applicable to this subpart
Appendix A—Reference documents

These sections are divided into four parts, which directly correlate with the four major parts of NFPA 70E. The simplification of OSHA 29CFR, Part 1910, Subpart S is indicated by the reduction in size of the document as compared to the 1971 National Electrical Code. This standard contains approximately 15,000 words, whereas the 1971 NEC contains approximately 250,000 words. Subpart S identifies the relevant portions of the NEC that are directly related to electrical safety in the workplace without the need for incorporating the entire NEC itself.

A controversy that has existed since the inception of OSHA concerns the retroactive clauses contained in the original standard. Many questions were proposed to the OSHA committee regarding the financial impact of this retroactivity. The original provisions regarding retroactivity were outlined in 1910.309(a), which applies to all electrical installations regardless of when they were built. The revised provisions relating to retroactivity are contained in Table 1910.302(b)(1). These provisions reflect changes from the old standard; they simplify and reduce the compliance burden and allow for alternative compliance methods. It was determined that the revised standard does not make a significant financial impact because it does not increase existing compliance requirements.

TABLE 1910.302(b)(1)—OSHA RETROACTIVE PROVISIONS

A list of the retroactive provisions taken from OSHA's Table 1910.302(b)(1) follows. The text for these references is taken directly from the National Electrical Code. The years that these sections first appeared in the National Electrical Code are listed. In addition, the original NEC section now may be referenced in another section in the current NEC edition; where this is done, appropriate present NEC references are provided in parentheses.

NEC Sections

110-14(a),(b)—Electric Connection—1930
110-17(a),(b),(c)—Guarding of Live Parts—1936
110-18—Arcing Parts—1937
110-21—Marking—1937
110-22—Identification—1965
240-16 (240-24)—Location in Premises for Overcurrent Protection Devices—1930
240-19(a),(b) (240-41)—Guarding of Arcing or Suddenly Moving Parts of Overcurrent Devices—1930
250-3(a),(b)—D.C. System Grounding—1930
250-5(a),(b),(c)—A.C. Systems and Circuits to Be Grounded—1930
250-42(a),(b),(c),(d)—Fixed Equipment Grounding—1930
250-44(a),(b),(c),(d),(e)—Non-electric Equipment Grounding—1930
250-45(a),(b),(c),(d)—Equipment Connected by Cord and Plug, Grounding—1930
250-50(a),(b)—Equipment Grounding Connections—1947
250-51—Effective Grounding—1937
250-57(a),(b)—Fixed Equipment, Method of Grounding—1937
400-3(a),(b) (400-7)—Flexible Cords and Cable, Uses—1930
400-4 (400-8)—Flexible Cords and Cable, Prohibited—1959
400-5 (400-9)—Flexible Cords and Cable, Splices—1947
400-9 (400-5) (240-4)—Overcurrent Protection and Ampacities of Flexible Cords—1937

NEC Articles

500—Hazardous Locations
501—Class I Installations—1925; 1928; 1930
502—Class II Installations
503—Class III Installations

This retroactive application has been shown to be reasonably necessary for employee safety, and it does not put an extreme cost burden on employers. It should be noted that the compliance requirements specified in this table of retroactive provisions are presented in a performance-oriented format.

MANDATORY OSHA REQUIREMENTS FOR ALL ELECTRICAL INSTALLATIONS

OSHA 1910.302(b)(1) indicates that the requirements contained in the following OSHA sections shall apply to all electrical installations and utilization equipment, regardless of when they were designed or installed.

Table of OSHA Mandatory Requirements

1910.303(b)—Examination, installation, and use of equipment
1910.303(c)—Splices
1910.303(d)—Arcing parts
1910.303(e)—Marking
1910.303(f)—Identification of disconnecting means
1910.303(g)(2)—Guarding of live parts
1910.304(e)(iv)—Location in or on premises
1910.304(e)(v)—Arcing or suddenly moving parts
1910.304(f)(1)(ii)—2 wire DC systems to be grounded
1910.304(f)(1)(iii)—AC systems under 50 volts to be grounded
1910.304(f)(1)(iv)—AC systems of 50 to 1000 volts to be grounded
1910.304(f)(1)(v)—AC systems of 50 to 1000 volts not required to be grounded
1910.304(f)(3)—Grounding connections
1910.304(f)(4) Grounding path
1910.304(f)(5)(iv)(a) through 1910.304(f)(5)(iv)(d)—Fixed equipment required to be grounded
1910.304(f)(5)(v)—Grounding equipment connected by cord and plug
1910.304(f)(5)(vi)—Grounding of nonelectric equipment
1910.304(f)(6)(i)—Methods of grounding fixed equipment
1910.305(g)(1)(i) and 1910.305(g)(i)(ii)—Flexible cords and cables, uses
1910.305(g)(2)(ii)—Flexible cords and cables, splices
1910.305(g)(2)(iii)—Pull at joints and terminals of flexible cords and cables
1901.307—Hazardous (classified) locations

In comparing the NEC sections and articles specified in Table 1910.302(b)(1) and the OSHA referenced sections, one finds some differences in wording and application. For instance, 1910.304(f)(4)—"Grounding Path" specifies that the path to ground from circuits, equipment, and enclosures shall be permanent and continuous. This differs from the "Effective Grounding Path" for circuits, equipment, and metal enclosures for conductors specified in NEC Section 250-51, which indicates that this grounding path must be:

1. Permanent and continuous.
2. Of ample capacity to carry any fault currents likely to be imposed on it.
3. Of sufficiently low impedance to limit the voltage to ground and facilitate the circuit protective devices in the circuit.

The second and third requirements of 250-51 are difficult if not impossible to determine. These grounding paths are often in parallel with an array of other metallic bodies such as structural steel, water pipes, sprinkler lines, metallic raceways, and so on. Attempts to measure the impedance of such a path even with appropriate instrumentation may lead to improper conclusions about the integrity of the conductive path. In fact, because of the multiplicity of parallel paths, a low-impedance path may be indicated by a test instrument with the grounding conductor completely disconnected. OSHA 1910.304 does specify that conductors be protected in accordance with their ability to safely conduct current. OSHA 1910.305 requires that wiring methods, components, and equipment for general use be approved for their purpose and use. It was determined that if the provisions of 1910.304 and 1910.305 are followed, these two performance requirements will be met. For this reason, the ampacity and impedance requirements of NEC 250-51 are not specified. This situation should not affect employee safety.

Another example of the differences between the wording of the NEC and the OSHA electrical standards is found in the reference concerning the methods of grounding fixed equipment and the relationship between NEC 250-58(a) and OSHA 1910.304(f)(6). Section 250-58(a) states that the use of structural metal frames of buildings as equipment grounding conductors is prohibited. Although the use of the structural metal frames of buildings as equipment grounding conductors was permitted in the 1971 NEC, it was decided to maintain this concept in the revised OSHA Standard only for electrical equipment installed prior to April 16, 1981.

OSHA 1910.302(b)(2) indicates that "every electric utilization system and all utilization equipment installed after March 15, 1972, and every major replacement, modification, repair, or rehabilitation, after March 15, 1972 of any part of any electric utilization system or utilization equipment installed before March 15, 1972, shall comply with the provisions of OSHA sections 1910.302 through 1910.308." A "major replacement, modification, repair, or rehabilitation" includes work similar to that involved when a new building is constructed, a new wing is built, or an entire floor is renovated. Therefore, virtually the entire OSHA standard is retroactive for installations or equipment made or installed since March 15, 1972, as well as for installations and equipment installed prior to this date if major modifications have been made.

Table 5-1. Electric Utilization Systems and Utilization Equipment Installed after April 16, 1981.

OSHA—29CFR, PART 1910, SUBPART S	CORRESPONDING NEC SECTIONS AND ARTICLES (1990)
1910.303(h)(4)(i)(ii): Entrance and access to workspace (over 600 volts)	Section 110-34(a)
1910.304(e)(1)(vi)(b): Circuit breakers operated vertically	Section 240-81
1910.304(e)(1)(vi)(c): Circuit breakers used as switches	Section 240 83(d)
1910.304(f)(7)(ii): Grounding of systems of 1000 volts or more supplying portable or mobile equipment	Sections 250-5(c) and 250-154
1910.305(J)(6)(ii)(b): Switching series capacitors over 600 volts	Section 460-24
1910.306(c)(2): Warning signs for elevators and escalators	Section 620-52(c)
1910.306(i): Electrically controlled irrigation machines	Sections 675-8 and 675-15
1910.306(J)(5): Ground-fault circuit interrupters for fountains	Section 680-51
1910.308(a)(1)(ii): Physical protection of conductors over 600 volts	Section 710-3
1910.308(c)(2): Marking of Class 2 and Class 3 power supplies	Section 725-34
1910.308(d): Fire protective signaling circuits	Article 760

There are other provisions of Subpart S that apply to electric utilization systems and utilization equipment installed after April 16, 1981. They include the references in Table 5.1.

CRITERIA FOR OSHA'S ACCEPTANCE

Finally, it should be noted that in OSHA 1910.399, which includes the definition of specific terms applicable to the standard, there is a description of when an installation or equipment is acceptable to the Assistant Secretary of Labor and, therefore, approved within the meaning of Subpart S.

First, it is approved if it is accepted, certified, listed, labeled, or otherwise determined to be safe by a nationally recognized testing organization such as Underwriters Laboratories Inc., Factory Mutual Engineering Corporation, or another organization.

Second, if the installation or equipment is not tested, listed, certified, accepted, or deemed to be safe by a nationally recognized testing organization, it can become acceptable if it is inspected or tested by another agency or the local authority having the responsibility for enforcing the occupational safety provisions of the National Electrical Code, and if the installation or equipment is found in compliance with the NEC as it applies to this Subpart S.

Third, if equipment is custom-made, then the manufacturer must determine the safe use of the equipment and provide specific test data relating to its safety. The employer must keep the test data and make them available to OSHA representatives on request.

SAFETY PROBLEMS RELATED TO ELECTRICAL SHOCK

In the 10-year period from 1971 to 1981, the National Center for Health Statistics reported approximately 1,000 accidental electrocutions each year in the United States, about 25 percent of them industry- or farm-related. Of course, the actual effects of an electric shock on an individual will depend on the circuit type, voltage, amperage, body impedance, and time of exposure (see Chapter 2). It has been determined that ventricular fibrillation (the cessation of the rhythmic pumping action of the heart) can occur at extremely low amperage. This condition is often fatal. Also, severe injuries due to deep internal burns may be the cause of delayed fatalities even though the shocking current does not pass through vital organs. The burns suffered in electrical accidents are typically of three types: electrical burns, arc burns, and thermal contact burns. Electrical burns are caused by the flow of current through body tissues. The burn itself may be only on the skin surface, or deeper layers of the skin may be affected. Arc burns are caused by the extremely high temperature of the electric arc, as high as 42,000°C, in close proximity to the body. Thermal contact burns are associated with the skin coming in contact with the hot surfaces of overheated electrical components.

Another cause of accidents is the indirect effect of an involuntary muscular reaction to electrical shock. Such a response can cause bone fractures, loss of limbs, bruises, or even death from falls or contact with moving machinery.

Electrical accidents generally are caused by three factors: unsafe work performance, working with unsafe equipment, and environmental conditions that contribute to making the workplace unsafe.

LOCKOUT–TAGOUT REGULATIONS

Probably the most common cause of electrically related accidents is the failure to deenergize electrical equipment during repair or inspection procedures. The *Electrical Safety-Related Work Practices—Lockout–Tagout* regulations were approved on August 1, 1990 and became effective on December 4, 1990. These rules are based on Part II of NFPA 70E, entitled *Electrical Safety Requirements for Employee Workplaces*. The provisions of this lockout–tagout procedure are administered by OSHA. It was hoped that compliance with these regulations would prevent an estimated 78 fatalities and 3,400 injuries each year. Employers were required to initiate safety training procedures as mandated by these rules by August 6, 1991 for unqualified employees in certain high-risk jobs. The rules are designed to differentiate between electricians and linemen who are trained to work with electrical equipment and others who only occasionally work on equipment.

See Chapter 2 for further discussion of the physiological response to electrical shock.

REFERENCES

Design Safety Standards for Electrical Systems. 29CFR Subpart S. Washington D.C.: U.S. Department of Labor.

Electrical Safety Requirements for Employee Workplaces. NFPA 70E 1988 Edition. Quincy, MA: The National Fire Protection Association.

6

Electrical Wiring and Equipment in Hazardous (Classified) Locations

Gregory Bierals

Many subjective factors must be considered to properly classify a hazardous area, and many of them often are overlooked by even the most experienced electrical personnel. The National Electrical Code offers guidance in specifying definitions of these hazardous areas in Articles 500-5, 500-6, and 500-7. In addition, Appendix A to the 1987 NEC, as referenced in Article 90-3, paragraph 4, lists NFPA publications that provide additional information to aid the classification process. Also, interested persons should contact the American Petroleum Institute for booklets of recommended practices (RP-500A, 500B, 500C) in the classification of special occupancies. Underwriters Laboratories, Factory Mutual Engineering Corporation, The Canadian Standards Association, Crouse-Hinds Co., Appleton Electric Co., and various insurance carriers offer a wealth of information on the subject.

It is important to recognize the many variables that affect the classification procedure. The definitions referenced in the NEC for Class I, Class II, and Class III locations should be used for definitive purposes only and not to make the final designation for an area classification. The information contained in this chapter provides a basis for assessing these variables.

DEFINITIONS

Class I, Division 1 Definition

According to Article 500-5(a) of the National Electrical Code, a Class I, Division 1 location is:

1. One in which flammable gases or vapors are or may be present in the air in quantities sufficient to produce explosive or ignitible mixtures under normal operating conditions;
2. Or one in which ignitible concentrations of such gases or vapors frequently may exist because of repair or maintenance operations or because of leakage;
3. Or one in which breakdown or faulty operation of equipment or processes might release ignitible concentrations of flammable gases or vapors, and might also cause simultaneous failure of electric equipment.

Normally a Class I, Division 1 area is considered the equivalent of an "open" system.

Class I, Division 2 Definition

In accordance with Article 500-5(b), a Class I, Division 2 location is:

1. One in which volatile flammable liquids or flammable gases are handled, processed, or used, but in which the liquids, vapors, or gases normally will be confined within closed containers or closed systems from which they can escape only in case of accidental rupture or breakdown of such containers or systems, or in case of abnormal operation of equipment;
2. Or one in which ignitible concentrations of gases or vapors normally are prevented by positive mechanical ventilation although it might become hazardous through failure or abnormal operation of the ventilating equipment;
3. Or one that is adjacent to a Class I, Division 1 location, and to which ignitible concentrations of gases or vapors might occasionally be communicated unless prevented by adequate positive pressure ventilation from a source of clean air, where this ventilation system has suitable safeguards against loss of the system.

Normally a Class I, Division 2 location is considered to be the equivalent of a "closed" system.

DIFFERENTIATION BETWEEN CLASS I DIVISIONS 1 AND 2: AN EXAMPLE OF THE CLASSIFICATION PROCESS

It seems fairly easy to differentiate between these "open" and "closed" systems, but this is where most of the errors in proper area classification are made. Consider an indoor area, that is, one with two or more walls and a roof, that contains a sealed piping system and possibly sealed vessels containing, for example, liquid styrene. One might consider this area to meet the criterion of a Class I, Division 2 location (500-5b) and might classify it as such. However, doing so could be a serious mistake if an abnormal condition developed. For example, with the accidental rupture of this piping system, the supposed Class I, Division 2 area could become Division 1 in a matter of minutes because of inadequate ventilation.

The proper classification of this space requires a thorough examination of styrene, a volatile liquid, to determine its flammable and explosive properties. Liquid styrene is a Class I flammable liquid that has a flash point of 90°F. If the liquid is exposed to ambient temperatures at or above 90°F, it begins to evaporate to form an ignitable vapor. Therefore, the first step must be to determine the normal ambient temperature of the area. If the temperature is typically at or above 90°F, one might consider this space to be a Division 1 location. If the precaution of providing adequate positive mechanical ventilation has been taken, then one should be able to classify this area as Division 2.

In establishing the proper classification for an area, it is of paramount importance that one know and understand the physical properties of all the involved substances. In this example, these would include the lower explosive limit (LEL) and the upper explosive limit (UEL) of styrene, which are 1.1 percent styrene (LEL) and 6.1 percent styrene (UEL) when it is mixed with air. The presence of a mixture of styrene vapor and air within these explosive limits along with an ignition source may result in an explosion. If the vapor concentration is below the LEL, the mixture is too lean to ignite; and if it is above the UEL, the mixture is too rich to ignite. The explosion pressures created during the ignition process will depend on the specific concentrations of vapor and air. For example, if the vapor concentration is at or slightly above the LEL, the explosion pressure generated will be relatively weak. The same is true if the vapor concentration is near the UEL. Maximum explosion pressures will develop when the vapor-to-air concentration is in the midrange, approximately 3.6 percent.

Vapor Density Considerations

Another important factor is the vapor density, which is the molecular weight of the vapor of gas when compared with the weight of air. Heavier-than-air vapors will tend to lie near the floor surface of a room or an area. A slight breeze or disturbance can spread this fuel over a very large area; so the ignition of the substance may occur at quite a distance from the vapor source. Lighter-than-air vapors or gases will tend to rise rapidly, and if unrestricted can disperse quite readily.

Ammonia, as it is released from a liquid form to a vapor, displays the characteristics of a heavier-than-air substance initially. Then, later, as the vapor begins to warm, it becomes lighter than air and rises. Acetylene gas has about the same molecular weight as that of air; so it tends to float in air. If a vapor or gas is heavier than air, then any sources of ignition occurring at or near the floor surface should be of concern.

Ignition Sources

A potential source of ignition includes any thermite reaction that may develop, as would occur if a ferrous tool were to strike a concrete floor, or if a piece of rusty steel struck a light alloy material such as aluminum. In the latter case, the thermite reaction is caused by iron oxide striking the light alloy material. Although ignition sources may have relatively low energy levels, they typically can produce enough heat to cause ignition. The minimum value of energy, measured in millijoules, required to cause the ignition of various substances is presented in Table 6.1. The table values are based on an equivalent capacitive spark discharge.

There also is a minimum ignition current, measured in milliamperes, that can cause an ignition process. The minimum ignition currents are based on an equivalent inductive spark discharge. See the listing of minimum ignition currents presented in Table 6.2.

Levels of ignition energy are measured by an instrument known as a break-flash device which can simulate both capacitive and inductive type spark dis-

Table 6-1. Ignition Energies.

Class I		
Acetylene and hydrogen	0.017	mJ
Cyclopropane	0.18	mJ
Ethylene	0.08	mJ
Methane	0.30	mJ
Propane	0.26	mJ
Class II		
Aluminum	15	mJ
Magnesium	40	mJ
Soft coal	30	mJ
Grain	240	mJ

Table 6-2. Minimum Ignition Currents.

Acetylene	60 mA
Ethyl ether	145 mA
Hydrogen	75 mA
Propane	146 mA
Methane	195 mA

charges. Probably the most accurate of these devices is the West Germany Break-Flash. This latest device is more accurate than earlier break-flash models. However, the minimum ignition energy and minimum ignition current levels specified in the tables are based on earlier and less accurate devices and they have become accepted as the standards of the industry.

Currently a test instrument, described in IEC Standards 79-3 and 79-4, is available to conduct spark ignition testing. This device consists of an explosion chamber with a volume of 15.25 cubic inches. Inside the chamber are two rotating discs separated by approximately 0.401 inches. The upper disc has 4 very fine tungsten wires attached. The lower disc is made of cadmium and contains two slots. The free length of the tungsten wires is 0.039 inches longer than the separation distance between the two discs. The upper disc rotates at 80 rpm and the lower cadmium disc rotates in the opposite direction at 19.2 rpm. The discs are insulated from one another and also from the housing of the instrument. One lead of the circuit under test is attached to the upper disc and the other lead is connected to the lower disc. A capacitive spark discharge is simulated as the tungsten wires make and break when they pass over the slots in the cadmium disc. During the test, the explosion chamber is filled with a representative fuel-air mixture. For direct current circuits, at least 400 revolutions of the tungsten wire holder are required, 200 revolutions for each polarity. Similarly, for alternating current circuits at least 1,000 revolutions are required. Typically a safety factor of 1.5 is used, i.e., the current or voltage is increased during the test to 1.5 times the value of normal conditions. Other fault conditions can also be tested such as a short-circuit or a ground-fault. No ignition of the fuel-air mixture is permitted for a circuit to successfully pass the test. This test apparatus is used internationally and is included in the International Electrical Commissions requirements for spark ignition testing.

Ventilation Requirements

Ventilation requirements are an important consideration in area classification. As discussed earlier, if positive mechanical ventilation is provided, an area may be considered as Division 2 if the area could become hazardous through failure or abnormal operation of the ventilating equipment. For a Class I liquid, the ventilation usually is considered adequate if it is equivalent to at least one cubic foot per minute for each square foot of solid floor area. If the ventilation is inadequate, and the ambient temperature may exceed the flash point of the liquid, then a rupture or a leak in a sealed piping system could cause a Division 2 location to become Division 1 in a very short period of time.

Surface Temperature Considerations

The ignition temperature of the substance should be known as well as the maximum surface temperatures of all equipment operating in the area. The National Electrical Code permits standard, open type, squirrel cage motors with no arcing or sliding contacts in a Division 2 area (NEC Article 501-8b). This equipment may appear to be a hazardous condition in itself, but it typically is not. As the motor operates, it is subject to a great deal of turbulence. Any concentrations of ignitible vapor or gas/air mixtures would flow across the surface of the open motor, and, in making contact with this surface, would heat and rise and very rapidly would be replaced by cooler vapor or gas/air mixtures. Thus these substances never would reach the surface temperature of the equipment, as long as the motor continued to operate. Of course, the motor should not be overloaded, so as not to develop excessive temperatures.

When a motor develops a stalled or locked rotor condition, the surface temperature of its housing can increase significantly. Article 501-8(b) of the NEC recognizes that in a typical Class I, Division 2 installation, the likelihood of a process or equipment failure releasing ignitible concentrations of fuel/air mixtures coincident with a stalled or locked rotor condition is remote. Therefore, the potential for a hazardous condition is not significantly increased by the installation of a standard, open-type, squirrel-cage, induction motor with no arcing or sliding contacts in lieu of a more expensive explosion-proof motor.

The installation of multiple on-line, open-type motors will, however, increase the overall probability of achieving a hazardous condition because each motor provides an additional potential source of ignition. There are procedures which can mitigate the potential of having such a hazardous situation occur. For example, one can obtain from the manufacturer the "locked rotor time rating" for each motor. Then a separate thermal overload device can be installed for each motor. These overloads should be installed in appropriate explosion-proof enclosures if in the Class I area, or in a general purpose enclosure if located in a safe unclassified area. Each thermal overload device should be carefully selected so that motor power is interrupted before the surface temperature of the motor housing approaches approximately 75 percent of the ignition temperature.

One can easily determine if the selected thermal overload provides adequate protection against excessive overheating of the motor. A comparison is made between the motor's heating curves and the heating curve of the thermal overload device. The system is safe as long as the overload de-energizes the motor well before fuel-air ignition temperatures are reached.

Once again, a hazardous situation can exist only if there is a simultaneous release of ignitible concentrations of fuel-air mixtures (an abnormal condition in a Class I, Division 2 area) and a stalled or locked-rotor condition. However, these situations can develop. A simple solution is to provide protection against developing excessive surface temperatures as described. This added protection circuitry should not be an economic burden, especially when one compares the price differential between using a standard open-type motor housing and an explosion-proof type motor.

CLASS I LOCATIONS

Explosion Proof Enclosures

An understanding of the design and usage of explosion proof enclosures is an important prerequisite to the installation of wiring systems in a Class I environment. These enclosures are specially designed to withstand at least four times the explosion pressures produced during ignition. The hot gases associated with this ignition process are cooled in three ways: (1) by the refrigeration effect—as the heated gases expand from a high pressure to a low pressure, they begin to cool naturally; (2) by absorption—some of the heat is absorbed by the enclosure itself where there is a definite temperature rise of approximately 5°C to 15°C; and (3) by turbulence (see below).

Thread Engagement Requirements for Conduits and Box Covers

The NEC requirement is for at least five full threads to be engaged on threaded conduit connections and threaded box covers in a Class I hazardous location (NEC Article 501-4a). The reason for specifying five full threads is to create a long enough flame path that the enclosure has sufficient time to absorb the heat from the ignition process. The flame produced during ignition is subjected to a tortuous path as it propagates around these threaded connections; flames do not circulate around corners very well. The third means of cooling is by turbulence: when the heated gases finally emerge from the enclosure, they mix with the surrounding air, and this rapid mixing process produces a natural cooling process.

Conduit and Cable Seals

Article 501-5a(1) specifies that a conduit or cable seal be installed within 18 inches of a box that contains arcing components or 18 inches from an enclosure if any conduit connections are 2-inch trade size or larger. The reason for the latter requirement is that as the volume of the conduit or enclosure increases, explosion pressures also increase. The 18-inch dimension is required because during ignition flames would travel through any connected conduits at a velocity exceeding the speed of sound. A sonic wave is produced that travels ahead of the flame front. When the sonic wave and the flame finally reach the compressed gases and detonation takes place, the explosion pressures created can be ten times above normal; and it has been found that this condition would be most hazardous in fairly short lengths of empty conduits. Fortunately, most conduits are filled to some degree with conductors, thus restricting the creation of this sonic wave. Also, longer conduit lengths provide some further restriction, as the sonic wave is subjected to friction along the walls of the raceway. The worst-case condition typically occurs where the conduit is relatively short, or about 5 to 10 feet long; so at 18 inches a conduit or cable seal provides a margin of safety.

CLASS II LOCATIONS

Dust-Ignition-Proof or Dust-Tight Enclosures

One common misconception about explosion-proof enclosures is that if they are suitable for Class I locations, then they automatically are suitable for Class II locations as well. This is not true unless the equipment also is identified for Class II environments. In fact, explosion-proof equipment is not even required for Class II applications. Typically, enclosures in these areas must only be dust-ignition-proof or dust-tight, depending on whether the area is classified as Division 1 or Division 2.

Dust-ignition-proof enclosures are designed to prevent the entry of ignitible amounts of dust and to keep arcs, sparks, or excessive heat generated inside the enclosure from igniting any exterior accumulations of a specific dust on or in the vicinity of the enclosure. It is interesting to note that ignitible concentrations of dust are given two different ignition temperatures. One value, is for a "dust cloud," and the other is for a "dust layer." In examining the "spontaneous ignition temperature" of various types of dust, one notices that in most cases the ignition temperature of a dust layer is lower than that for a dust cloud. Some examples are magnesium, zinc, bronze, chromium, tin, cadmium, coal, starch, cocoa, corn, wheat, malt, and alfalfa. The reason for this is that dust that has settled on enclosures and equipment can become very dry, and dust that is carbonized or dry can become highly susceptible to spontaneous ignition. In this situation no arc or spark is required for ignition.

Dust-ignition-proof enclosures are required for switches, circuit breakers, motor controllers, and fuses in Class II, Division 1 locations, whereas these same devices only would require enclosures that are dusttight in Class II, Division 2 locations. Dusttight enclosures are constructed so that dust will not enter the enclosing case under "specified test conditions." These test conditions are listed in ANSI/NEMA 250-1985, Clause 6.5, "Enclosures for Electrical Equipment, 1000 Volts Maximum." The ignition energy for a specific dust is based on concentrations of the dust measured in "ounces per cubic foot." Typically, to ignite a cloud of dust there must be a sufficient concentration of it in suspension. Usually this concentration is enough to obscure one's vision. In other words, you would have difficulty seeing your hand in front of your face.

Dust Explosion Pressures

The explosion pressures created by dust as it ignites are not too dissimilar, in many cases, to those created in Class I applications. For instance, the explosion pressure of magnesium is 90 psi, and that for hydrogen is 102 psi. However, zinc, bronze, and chromium produce only about half the explosion pressure of magnesium.

One important consideration is that in areas containing metallic dusts there are only Division 1 locations and not any Division 2 locations. (See NEC, Article 500-6a). The reason for this restriction is twofold:

1. Dusts of an electrically conductive nature (metallic dusts) can enter enclosures much more easily than other types of dusts.
2. Because of their conducting abilities, it is quite easy to develop conducting paths from conductors inside an enclosure to ground or possibly to adjacent conductors. Arcing or sparking conditions that may develop inside an enclosure certainly can produce an ignition source.

INTRINSICALLY SAFE CIRCUITS AND EQUIPMENT

The concept of intrinsic safety in hazardous (classified) locations is not a new one. The National Electrical Code did not contain a reference associated with this type of system until 1968. The Canadian Electrical Code was devoid of such a reference until 1966. However, intrinsically safe circuits and equipment were in use in Great Britain as early as 1917. This equipment consisted of a transformer and bell circuit used as a crude means of communication in a coal mine. If a certain type of bell was matched with a specific type of transformer, the annunciator circuit was not capable of causing ignition of any combustible gases or ignitible dusts present in that environment. This early safety device was developed because of two disastrous coal mine explosions, thought to have been caused by a common bell circuit, similar to that used for an ordinary door bell. To prevent ignition, a pair of wires were run through the mine, separated by a distance of approximately 4 inches. These wires were short-circuited by squeezing them together or possibly by a piece of metal. The circuit operated at 12 V and was power-limited by the resistance of the secondary windings of the transformer. However, the ignition energy of the equipment and circuit was still above that required to ignite the gases or dusts present.

Following the second explosion in 1913, a study was made about system components that could be safely used in this environment. The resultant bell signaling circuit, first used in 1917, became the precursor of the modern intrinsically safe systems. The transformer used for this circuit was an early form of the zener barriers used today. It was not an intrinsically safe device and could not be installed in a hazardous location.

Organizations such as Underwriters Laboratories and the Canadian Standards Association did not test and identify intrinsically safe equipment until the 1950s. The dictionary defines the word *intrinsic* as pertaining to the essential nature of a thing; it comes from the Latin word *intrinsecus,* which means inwardly, on the inside. The concept of intrinsic safety as applied to wiring methods in hazardous (classified) locations is to provide a circuit and components that are incapable of causing ignition of a flammable or combustible vapor, gas, or dust in its most easily ignitable state. This type of system will not release sufficient electrical or thermal energy to cause ignition of these substances under either a normal or an abnormal condition. These conditions may be the result of a short circuit, a ground fault, an open circuit, or overvoltage.

It should be noted that installing an intrinsically safe system does not necessarily guarantee a safe system. When such a circuit and equipment are tested, there are simultaneous fault conditions that must be considered. The system must remain safe when subjected to these faults.

NONINCENDIVE EQUIPMENT AND CIRCUITS

There is another type of system referenced in NEC, Section 501-3(b)(1), exception C, which is identified as *nonincendive*. Its concept is similar to that of the intrinsically safe system, in that such a circuit or component also would be incapable of releasing sufficient electrical energy during normal conditions to cause ignition. The abnormal conditions are the same as mentioned before, that is, a short circuit, a ground fault, an open circuit, or overvoltage. However, the major difference between these systems is that thermal energy under abnormal conditions of operation is not considered for nonincendive circuits and components as it is for intrinsically safe systems. For this reason, nonincendive concepts are acceptable only for Division 2 applications, whereas intrinsically safe systems and equipment may be used in Division 1 or in Division 2, depending on the listing of the particular system. By their very nature, nonincendive systems are low energy systems. A nonincendive system may consist of a nonincendive circuit, nonincendive contacts, nonincendive components, and nonincendive equipment. However, nonincendive contacts do not necessarily make a nonincendive circuit. The contacts may be nonincendive, but the circuit still can release above-ignition-level energy if subjected to a fault condition such as an open circuit. Low energy contacts contained in a very small housing may constitute an example of a nonincendive component. Nonincendive equipment may be battery- or solar cell–operated. Many hand-held calculators, paging receivers, and battery-power watches are examples of this type of equipment. The testing and evaluation procedures for nonincendive circuits and equipment are virtually the same as for intrinsically safe circuits and equipment. The use of nonincendive equipment in any hazardous (classified) location can pose a problem because if the equipment is portable and is dropped, fault-producing ignition-capable energy may result.

EXAMINATION AND APPLICATION OF INTRINSICALLY SAFE CIRCUITS AND COMPONENTS

Article 504 of the NEC covers the installation of intrinsically safe apparatus and wiring for Class I, II, and III locations. For further information regarding the installation of instrument systems in hazardous locations, see *Installation of Intrinsically Safe Instrument Systems in Class I Hazardous Locations,* ANSI/ ISA RP 12.6, 1987. Information on test conditions concerning intrinsically safe apparatus are described in ANSI/UL 913, 1988. Ignition energy levels of Class I and Class II vapors, gases, and dusts are quite low. Some examples of these energy values, expressed in millijoules, are listed in Table 6.1. These values are known as the minimum ignition energies for these substances (see above, under "Ignition Sources").

The testing of intrinsically safe apparatus and related circuits must take these low energies into consideration. (The test instrumentation used to simulate capacitive discharge ignition energies was discussed earlier, under "Ignition Sources.") Inductive spark discharges (measured in milliamperes) also are possible and are quantified by appropriate tests to determine the minimum ignition current for a particular substance. Some representative values are presented in Table 6.2.

Design of Intrinsically Safe Circuits

The design of an intrinsically safe circuit needs to consider both capacitance and inductance values; the cable inductance and the capacitance of the circuit loop must be known. Detailed information regarding cable inductance and capacitance is available from the manufacturers of intrinsically safe equipment.

With respect to the three legs of an explosion or fire triangle (fuel, oxygen, and heat), purging methods (see NFPA 496) block the fuel source, pressurizing an enclosure may block the fuel and/or the oxygen source, explosion-proof enclosures block the heat source, and intrinsically safe circuits and apparatus prevent the heat source from being an ignition hazard.

Any safe system that is built must be of such low energy that even when it is subjected to two simultaneous faults, the energy release is still incapable of causing ignition of a substance in its most easily ignitable state. This is achieved by using a device known as a zener current or zener voltage barrier. (See Figure 6.1.) The device may operate at a voltage of up to 250 V on the supply side, whereas voltage levels normally are limited to no more than 30 V on the protected side although some barriers have a voltage as high as 100 V for Group C and Group D applications. The barrier typically consists of a series of resistors, two zener diodes, anad a fuse encapsulated into a single unit. Components are installed in multiple so that in the event of a single component failure, the system remains safe. The barriers are usually of the 4- to 20-mA variety. If the circuit and component resistances are known, the system voltage can easily be determined. If a condition that produces an overvoltage develops, such as the failure of a resistor, the zener diodes provide a low-impedance current path to ground to maintain the safety of the system. If one diode fails, the other will provide the safe circuit path to ground. The barrier itself is connected to two copper ground bus bars that are connected to a grounding electrode. The resistance to ground of this system must be quite low, on the order of one ohm or less. The grounded loop conductor is referenced to ground only at the barrier and not downstream because of the potential differences and objectionable currents that may result if multiple connections to ground are made.

NEC Section 504-30(a) requires separation of the intrinsically safe loop wiring from all other wiring systems so that a minimum of a 2-inch separation is maintained in common enclosures (such as the barrier enclosure). The intrinsically safe wiring is to be installed in separate raceways or possibly separated by a fixed barrier, if a common raceway is used, in order to maintain system safety. Section 504-30(b) requires the wiring of one intrinsically safe system to be kept separate from another intrinsically safe system by the installation of separate

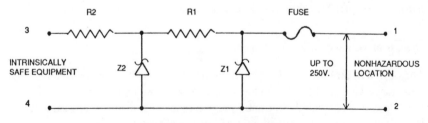

Figure 6.1. Fused zener barrier.

cables or raceways, or, if in common raceways, by insulation of at least 0.01 inch thickness.

Section 504-70 specifies that when wiring from an intrinsically safe system passes from a hazardous location to a less hazardous location, proper cable sealing techniques must be used. This is especially important for a pressurized enclosure containing intrinsically safe components even though the cable itself has a gas/vapor-tight outer jacket on the outside of the enclosure. Because of the pressure differential across the cable, small amounts of potentially explosive gases or vapors could be transmitted through the cable core to these areas.

Section 504-80(b) requires the identification of all raceways, cable trays, and open wiring used for intrinsically safe system wiring. Permanently affixed labels are to be applied at not more than 25-foot intervals with the wording "Intrinsic Safety Wiring" or its equivalent. The individual conductors may be identified by a light-blue-colored insulation, provided that no other conductors of other systems also are colored blue. The terminals and junction locations also are to be identified as a part of an intrinsically safe system in accordance with Section 504-80(a).

REFERENCES

Electrical Construction Materials Directory, 1989 (Green Book). Northbrook, IL: Underwriters Laboratories.

Electrical Installation in Hazardous Locations by P. J. Schram and M. W. Earley, 1988, Quincy, MA: National Fire Protection Association.

General Information for Electrical Construction, Hazardous Location, and Electric Heating and Air Conditioning Equipment, 1987 (White Book). Northbrook, IL Underwriters Laboratories.

Guide for Use of Electrical Products in Hazardous Locations. 1990 Appleton NEC Code Review. Chicago, IL: Appleton Electric Co.

Hazardous (Classified) Locations. 29CFR Part 1910.307. Washington, D.C.: U.S. Department of Labor.

Recommended Practices for Classification of Locations for Electrical Installations in Petroleum Refineries. RP-500A, Fourth Edition. Washington, D.C.: American Petroleum Institute.

Recommended Practices for Classification for Electrical Installations at Drilling Rigs and Production Facilities on Land and on Marine Fixed and Mobile Platforms. RP-500B, Fourth Edition. Washington, D.C.: American Petroleum Institute.

Recommended Practices for Classification for Electrical Installations at Pipeline Transportation Facilities. RP-500C, Fourth Edition. Washington, D.C.: American Petroleum Institute.

7

Electrical Fires: Causes, Prevention, and Investigation

Theodore Bernstein

An understanding of the electrical equipment and systems involved is essential when one is investigating the cause of a fire alleged to be of electrical origin. Even if electricity was the only known ignition source in the alleged area of fire origin, it does not necessarily follow that an electrical fault caused the fire. To prove that electricity caused the fire, one first must show how an electrical fault ignited a fuel source and then how the fire spread from the region of the fault. It cannot be assumed that an electrical component started the fire because it seemed to be the only possible cause of the fire in the area of fire origin. There may have been an error in determining the area of origin; there may have been something unusual or unexpected in the area of origin; or an unusual set of circumstances may have occurred. For example, a bag of apples had been placed on a kitchen counter in front of a toaster oven. The apples in the bag shifted, tipping the bag over and turning on the oven. Then the bag containing the apples started to burn. Because the home was equipped with a sprinkler system, the fire was quickly extinguished, and the sequence of events leading to the fire was easily reconstructed. Without the sprinkler system, the home might have burned; and because the toaster oven was the only ignition source in the area, it might have been mistakenly identified as the cause of the fire.

Finding arced conductors or electrical equipment with electrical faults after a fire certainly is not sufficient proof that the fire was of electrical origin; such fire-caused damage is commonly found after a fire. For any fire, as long as the electrical system was energized at the time of the fire, electrically faulted conductors or electrically damaged equipment can be expected. Thus great care must be exercised in correctly identifying the cause of a fire. One must not attribute a fire's origin to electricity just because electrical components were present. Electricity does cause fires but a competent fire investigator must still prove what equipment failed, what fuel burned, and how the fire spread from the faulted equipment.

CONDUCTORS AND FIRES

Insulation

The insulation requirement for an electrical conductor is determined by its service conditions: the voltage of the system, the ambient temperature and environment surrounding the conductor, and the electrical continuous current the wire is expected to carry. Higher voltages require thicker or more effective insulation. The life expectancy for the insulation, and consequently the equipment in which it is used, depends on the temperature history of the insulation. A decrease in the life span or even failure will result from excessive elevated temperatures, even for short time periods. Recommended types of insulation for various environments are listed in Table 310-13 of the National Electrical Code (1990). The ambient temperature, the temperature rise produced by the resistance heating of the current-carrying conductor, and the cooling heat transfer to the surroundings all must be considered in determining the maximum operating temperature of the insulation. The continuous current rating or ampacity for an insulated conductor is determined by the following factors: (1) the temperature rating of the insulation, (2) the heat generated by the steady state continuous current, and (3) the heat transfer to the environment permitted by the ambient temperature. Under continuous operation, heat generated by current in the conductor must not increase its operating temperature above the temperature rating of the insulation.

For example, a conductor with an insulation rating of 75°C placed in an operating environment of 75°C has zero current-carrying capacity. See Tables 310-16 through 310-19 of the National Electrical Code (1990) for the ampacity of wire and for different configurations of conductors.

A commonly used insulation for electrical wires is the plastic polyvinyl chloride (PVC). This insulation can have a maximum operating temperature of 90°C or 194°F and will start producing hydrochloric acid at a temperature of 115°C or 239°F (Sillars 1973). PVC decomposes at 200°C to 300°C or 392°F to 572°F (Hilado 1982). Its relatively low melting temperature explains why the insulation on wires usually is missing when conductors have been in the area of a fire. Polyvinyl chloride plastic insulation used for wire insulation will burn when exposed to a flame, but it will self-extinguish if the flame is removed. A flame will not propagate along a conductor if it is applied to the insulation at one point on the conductor.

Melting of Conductors in Fires

Aluminum often is found melted after a fire, copper infrequently is found melted, and steel almost never is found melted. This phenomenon can be explained by noting the melting temperatures for these materials:

Aluminum: 660°C or approximately 1,200°F
Copper: 1,083°C or approximately 2,000°F
Steel: 1,535°C or approximately 2,800°F

An electric arc, however, with temperatures in the range of 2,000°C to 4,000°C for a 2 to 20 A current (with even higher temperatures at higher currents) can melt any of these metals. There must be sufficient energy in the arc to melt the metal mass, as it is the heat content of the arc and not the temperature alone that causes the melting. A low-current spark will have a high temperature, but will lack the energy required to raise the temperature of a metal to its melting point.

A copper conductor can melt at a temperature below the melting point of pure copper if molten aluminum drips on the copper during a fire, producing a copper–aluminum alloy with a melting point below that of pure copper. Copper–aluminum alloys with more than 40 percent aluminum will have a melting point lower than that for pure aluminum. Such melted alloyed copper can give the appearance of being melted by arcing (Beland, Roy, and Tremblay 1983; Beland 1984).

Thin-gauge steel may exhibit holes that appear to be caused by melting in a fire. However, such holes usually are due to the rapid oxidation of the steel when subjected to a fire with temperatures lower than the melting point of steel.

Copper conductors, when exposed to temperatures in a fire below their melting point, often will anneal and lose their elasticity and strength. After a fire such copper conductors often are found to be of greatly reduced strength, crumbling in the investigator's hand or draped between wall studs or ceiling joists.

Conductors with Nicks or Broken Strands

Sometimes investigators point to a nick in a solid conductor or to some parted strands of a stranded conductor as a source of overheating. They attempt to explain how a conductor carrying a current at or below its rated ampacity can overheat and cause a fire, arguing that the reduced cross-sectional area at the nick or parted strands caused an overheating condition. However, it has been shown that the ampacity (see Chapter 3) of a conductor is determined by the allowable temperature rise of the conductor that is safe for the insulation. The analysis that was used assumed a uniform current density and power dissipation along the entire length of the conductor. The current density and power loss will be increased at the nick or parted strands. However, the temperature increase at the damage location will be minimal because any increased heat that is produced will be conducted away to the rest of the conductor mass, so that the temperature rise at the reduced cross-section will be limited (Bernstein 1983).

Aluminum Conductors

Aluminum conductors carry current just as well as copper conductors, and any problems with aluminum conductors usually occur at a terminal or a connection point. If the terminals or connections are not correctly chosen for the application or are not correctly installed (UL listed or labeled for the specific application), the aluminum can cold-flow. If that happens, a tight terminal or connection will loosen over time. The loose terminal or connection can overheat and cause equipment damage or a fire, even when carrying rated current or less. The gradual loosening of terminals is produced by expansion and contraction of the aluminum conductor caused by I^2R heating and cooling as the current is turned on and off. In addition, any oxide on the aluminum can increase the contact resistance of the terminal and lead to increased I^2R heating at the connection.

To ensure a safe installation, one must properly clean aluminum conductors and terminals and apply the correct pressure or torque to the terminals or connectors. One must be careful to make aluminum installations in strict accordance with the wire or terminal manufacturer's instructions.

ELECTRIC ARCS CAUSING FIRES AND CAUSED BY FIRES

Arc Initiation

The breakdown strength for air that can initiate an arc depends on the shape of the electrodes and the waveform of the applied voltage. A typical value is 30 kV per cm or 76.2 kV per inch; that is, there must be a voltage difference of 30,000 V for each centimeter of gap length or 76,200 V for each inch of gap length to initiate an arc through air. It is a fundamental property of arcs that no matter how small the air gap between conductors, there can be no breakdown across an air gap if the voltage is below approximately 300 V; but once an arc is initiated, only 20 V per cm or 50.8 V per inch is required to maintain the arc if there is sufficient electrical energy. Thus, once started, an arc can be maintained over a much longer air gap length than the original initiation distance. The arc voltage after breakdown is practically independent of the arc current.

Electric Arcs Caused by Fires

Fire will char or melt insulation, and the voltage difference between conductors can initiate an arc at voltages of 120 V or less, even though the conductors are not touching. The hot conducting gases or charred insulation provides the fault current path to initiate the arc. Because the conductors are not touching during the arcing, there will be a voltage drop in the arc that causes the fault current in the arc to be smaller than it would be if the conductors were in direct contact. This will increase the time necessary to trip any circuit overcurrent protection device and may allow considerable arcing to occur during a fire before any overcurrent protection device opens the circuit. The overcurrent protection may never operate, as the arc may burn itself clear. If the conductors touch, large fault currents will occur, and the overcurrent protection device will open quickly, tending to limit the arcing at the point of touching (Beland 1981).

Arcs can be initiated at lower voltages, well below the 300 V previously mentioned. They can occur when a low-resistance path develops over or through damaged insulation, allowing current to char the insulation and cause carbon tracking. As the insulation is further damaged, the current increases, and an arc can be initiated. Voltages as low as 6 V, such as from an automotive battery, can produce arcs as long as sufficient current is available from the power source to support the arc.

Electric arc welders have an 80 V open circuit voltage and a constant current power supply. Arcs are initiated either by making contact with the electrodes and separating them or by utilizing a high frequency, high voltage, low current power supply that produces a spark in the air gap between electrodes to initiate the arc.

Considerable arcing can be found after a fire when energized electrical equipment insulation is damaged. This will allow arcing across fire-damaged, charred insulation or through hot, conducting gases. Because of the intrinsic resistance of the arc column, the arcing current is limited, and considerable damage may occur before the circuit protection device operates. The current in the arc, even though it may be lower than the ampacity of the conductors in the system, can cause damage at the arc's terminus points, as the heat produced by an arc is greatest at these points. The current flow in the conductors away from the arc would do little, if any, damage to the conductors, as the magnitude of the arc current may be even less than the ampacity of the conductor.

Both of the terminus or attachment points for an arc should be identified. Apparent arcing damage found on only one copper conductor may indicate that the damage was not caused by arcing but by copper–aluminum alloying and melting.

Fires Caused by Electric Arcs

When multiple locations of arcing damage are found, the investigator must be able to show why a particular set of arc damage was the origin of the fire. When only a single instance of arcing damage is found in the area of fire origin, it must be shown how or why this arcing initiated the fire instead of occurring as a result of the fire.

Electrical systems in homes and industry are essentially constant-voltage, low-impedance systems; so if there is a conducting fault across two conductors in the system, the fault current magnitude will be determined primarily by the resistance of the fault. When two conductors touch, the fault current may be very large; on the other hand, the current in an arcing fault may be large, but it is certainly less than it would be if the conductors were touching. For low voltage systems, under 600 V, there will be essentially no arcing through the air between the conductors. If the conductor insulation is not charred or melted by a fire, the only ways to initiate arcing are through touching and parting of the conductors or failure of the conductor insulation, allowing arc tracking across or through the insulation.

Prior to a fire, limited arcing can occur if insulation is damaged, and two bare conductors touch and then part. This is not a common occurrence because when the conductors touch, the fault current in this low-impedance system will be high, and the circuit overcurrent protection device will operate rapidly to deenergize the system. The resulting damage in this situation usually is minimal. One also must consider what caused the conductor movement and contact.

Some investigators attempt to compare arcing associated with household or industrial electrical systems to arcs produced by electrical arc welders, but this is an inappropriate analogy. An electric arc welder represents a constant-current supply and not the constant-voltage supply found in home and industrial electrical systems supplied by an electric utility. When welding electrodes are touched to initiate an arc, the current is limited to a preset value, which will be the current in the arc after the electrodes are parted and an arc is struck. In constant-voltage systems, the currents in an arc are not controlled or limited and can become quite large. Thus, circuit overcurrent protection devices usually will operate and deenergize the circuit. It is virtually impossible to weld using a constant-voltage supply such as that provided by a utility. The usual 120 V, 60 Hz, constant-voltage utility source makes a very poor welding system because of uncontrolled, unlimited wide fluctuations in current.

When there is damaged insulation and the conductors in a low voltage system are not touching, arcing can be initiated by a charring and breakdown of the insulation. Merely scraping the insulation off a conductor will not necessarily provide a surface for arcing to track across. In fire investigations where arcs are alleged to have tracked over insulation, the contaminant that allowed tracking should be identified.

HEATING OF CONDUCTORS BY TRANSIENT FAULT CURRENTS

In electrical fire investigation cases, there sometimes are questions about the opening or parting of electrical conductors. Separation of conductors can be caused by mechanical stress, external fire, arcing to adjacent metallic structure, or overheating and subsequent fusing or parting of the conductor due to excessive current. This section describes a technique for determining the temperature rise in a conductor caused by transient currents before the wire has reached the final, steady state temperature. A methodology is described for calculating transient temperature rises for copper, aluminum, and steel conductors. Copper and aluminum are the usual conductor materials, but steel cables are used for overhead lightning shield wires, guy wires, and other structural supports.

The temperature rise for a current-carrying conductor can be calculated for the following conditions:

- The steady state current condition.
- The transient current condition.

Steady State Conductor Temperature Calculations

To calculate the steady state temperature for a given current, the input electrical energy converted into heat in the conductor is set equal to the sum of the three cooling output mechanisms—the conductive, the convective, and the radiation energy outputs. A steady state temperature is reached when there is an equilibrium established between the input electrical energy converted to heat and the sum of the three cooling energy outputs. Such a steady state temperature can be calculated relatively easily (Bernstein and Chang 1969).

Transient Current–Temperature Rise Characteristics for a Conductor

When the electrical energy input to a conductor lasts for only a few seconds or a fraction of a second, a steady state thermal condition is never reached. Therefore, in this situation a transient temperature rise must be calculated.

The convective, conductive, and radiative cooling contributions are assumed negligible in calculating the temperature rise of a conductor for a short time

Table 7.1. Relationship Between Inverse of Current Density and Fusing Times.

MATERIAL	INVERSE CURRENT DENSITY FOR FUSING (cmil/A)
Copper	6.87/t
Aluminum	12.02/t
Steel	18.44/t

Table is valid only for time periods on the order of a few seconds; t is the fusing time in seconds.

period. Only the conductor's heat capacity is considered in calculating its short-term, temperature rise.

Barnes (1966) provided a technique for evaluating the current–time relationship for a given conductor temperature rise under transient conditions. The heat developed per centimeter length of conductor is found from the electrical energy input using the average resistance of the conductor as the value of resistance at half of the temperature rise. The electrical energy input is a function of the power input and the time duration of this input. For a short period, cooling mechanisms are neglected, and the I^2R heat produced by the current is absorbed completely by the conductor. The temperature rise of the conductor for such a transient condition is proportional to the heat energy input from the current and inversely proportional to the specific gravity, specific heat, and volume of the conductor.

Using the Barnes technique, the relationship between the inverse of the current density and the time to fuse or part a conductor is given in Table 7.1. It is important to note that the current density for fusing is valid only for short periods of time when the cooling mechanisms are negligible. This is evident because for long periods of time the values given by Table 7.1 indicate that even tiny values of current would cause the conductor to fuse. The relationships given in Table 7.1 are not valid for longer times because the cooling effects have been neglected in the calculation; the relationship is valid only for times on the order of a few seconds.

Experimental Results

In order to compare the theoretical value for the current density–time relationship to melt conductors to actual results, tests were conducted on steel, copper, and aluminum conductors. For these tests, a constant, 60 Hz, alternating current was maintained in a one-foot length of conductor that was at an initial ambient temperature of 20°C. The time required for the conductor to part then was measured. For these tests, three different wire sizes were used, although not all wire sizes were used for each conductor material. The bare, solid wire sizes used were:

AWG 8: Area = 16,510 circular mils
AWG 12: Area = 6,530 circular mils
AWG 14: Area = 4,110 circular mils

The results are tabulated in Table 7.2, where the average measured time to part for a given current is compared to the calculated value determined by using the relationship in Table 7.1. The error of the calculated time for the wire to melt with respect to the actual measured value of the time is included in the table.

The results shown in Table 7.2 indicate that the calculated values are conservative, in that the actual time to part a wire for a given current is longer than the calculated value. This is to be expected because the calculated value neglects any cooling effects on the wire from external sources. The results for copper show clearly that at lower currents, where the time to part is longer, the calculated values for time to fuse depart most from the actual value because of the external cooling of the wire.

When insulated wires were used, it was found that the measured time to part was somewhat longer and more variable than with bare wires. This was to be

Table 7.2. Time of Seperation for a Conductor Carrying a Constant 60 Hz Current.

MATERIAL	WIRE SIZE (AWG)	CURRENT (A rms)	INVERSE CURRENT DENISTY (cmil/A)	MEASURED TIME TO PART (s)	CALCULATED TIME TO PART (s)	ERROR FOR CALCULATED TIME (%)
Steel	12	200	32.7	4.60	3.13	32.0
	12	100	65.3	15.29	12.54	18.0
	8	500	33.0	4.26	3.21	24.7
Copper	14	400	10.3	2.84	2.24	21.1
	8	1,500	11.0	2.92	2.57	12.0
	8	1,250	13.2	4.35	3.70	14.9
	8	1,000	16.5	6.75	5.78	14.4
	8	800	20.6	11.04	9.03	18.2
	8	600	27.5	21.45	16.05	25.2
	8	500	33.0	31.78	23.00	27.3
Aluminum	8	1,000	16.5	2.53	1.89	25.3
	8	500	33.0	9.79	7.55	22.9

expected, as the insulation tended to preserve the molten metal contact until the insulation melted.

INVESTIGATION OF A FIRE SCENE

Investigators of alleged electrical fires should visit the fire scene immediately if possible. Often it is not possible to do so because the fire occurred long before the investigation, or the fire scene is so disturbed that a visit would prove useless. Even if the scene cannot be visited, a suitable conclusion may be possible if photographs, reports, and equipment can be examined at a later date. This section will discuss the type of findings the investigator often encounters at a fire scene.

"V" Pattern Burns

Burn patterns that produce a "V" pattern on a wall sometimes are used to indicate an area of fire origin where there may have been electrical equipment, the argument being that the electrical equipment started the fire at the point of the "V," and the fire expanded as it burned up the wall. Such evidence may be helpful in evaluating the possible area of origin, but such burn patterns also can occur when the fire originated elsewhere, and the "V" pattern was produced when burning debris fell in the area where the point of the "V" was located, or when there happened to be something combustible at that point that was ignited by a fire in the area. Sometimes the flow of hot air and gases from a fire will produce a "V" pattern in an area—quite often near corners—well away from where the fire started. Any electrical equipment located at the alleged area of fire origin, at the point of the "V," must be examined carefully to see if it shows any faults that could explain how a fire could have started. Additionally, any adjacent fuel source that could be ignited by the electrical fault must be identified. Thus, one must study "V" burn patterns with great care before deciding that the burn pattern does or does not point to the area of fire origin.

Arced Conductors

Conductors that show some evidence of arcing often are discovered after a fire. Usually they are found in a fire of any cause if the fire damages the electrical insulation of energized conductors and arcing occurs. A lack of arcing on conductors with severely damaged insulation would indicate that the conductors were not energized when the insulation was damaged by the fire.

The location of an arced conductor sometimes is a clue to the site of a fire's origin. Consider the following example: (1) arced conductors were found only in the area of a service entrance; (2) areas well away from the service entrance, however, clearly were involved in the fire. Such evidence might be an indication that the fire originated near the service entrance; then early in the fire the conductors at or near the service entrance were damaged. This caused a parting of the conductors or operation of the circuit overcurrent device. The early disconnection of power to the rest of the building prevented additional arcing as the fire spread into other areas.

Similar reasoning can be used for an individual branch circuit where arcing has occurred at widely separated locations. In such a case, the arcing farthest from the power source probably occurred first if arcing closer to the power source caused conductors to separate. As with all such clues, the arcing is only one of many factors that must be evaluated in the investigation of a fire scene.

Multiple arcing sites along a branch circuit conductor run in a room usually are caused by a fire impinging on the conductors. The fact that a branch circuit conductor has parted in several places sometimes is used to support the opinion that the fire must have progressed in a direction toward the power source. A fire progressing away from the power source would cause the conductors to arc and separate closest to the power source first. Power interruption would prevent additional arcing on the load side of the branch circuit. However, this is not necessarily true. In an intense fire, arcing can occur more or less simultaneously along the entire branch circuit conductor run.

Electrical fires rarely are started because of multiple arcing sites along a continuous run of insulated conductors. Such an arcing pattern usually is caused by a fire.

Two conductors usually must touch before arcing can occur in circuits with voltages under 600 V even if the conductor insulation is damaged at some point along a continuous run. If the conductors touch, large fault currents will flow with minimal arcing at the point of contact, and the circuit overcurrent protection device will operate. There is little evidence of arcing at the point of contact.

In order to have a fire caused by arcing along a continuous run of conductors, it is necessary to have both conductor insulation damage and a conducting path. The conducting path may be a metal staple or conductive contamination of the damaged insulation. For arcing to occur between two insulated conductors, each must have its insulation damaged at the same location. Plain water usually will not provide suitable conductive contamination to initiate an arc. A metal staple may damage the conductor insulation without necessarily leading to an arcing fault. In a fire, the charred insulation and the hot gases can provide ample contamination of the insulation to produce an arcing fault.

The appearance of arced conductors at the location of the arc cannot be used to indicate whether the arc was caused by the fire or caused the fire. The shape of the bead, ball, or point usually tells the investigator nothing as far as this question is concerned.

Location of Electrical Fire Origins

Electrical fires tend to originate at outlets or in equipment supplied from the outlet. For this discussion, the definition of an outlet is that found in the National Electrical Code (1990) (Article 100):

Outlet: A point on the wiring system at which current is taken to supply utilization equipment.

Fires may start in receptacles, outlet boxes, cord-connected appliances, ceiling lights, and other such devices supplied from the outlet. These are much more likely places for the origin of fires than locations along a continuous run of conductors.

Sometimes a fire of electrical cause allegedly will have originated away from the location of the equipment that supposedly caused the fire. One must be very careful about accepting such an allegation. Conductor insulation will not burn except in a fire; so overheated electrical equipment cannot ignite the insulation of a conductor. Insulation will not act like a candle and carry fire to another area with ignitable combustibles.

Large fault currents may overheat terminals or splices closer to the power source than the fault location, but operation of the circuit overcurrent protection device may limit any such overheating. High-resistance faults, by definition, will have low fault currents. Therefore, they should not overheat the conductor insulation, terminals, or splices in the circuit upstream from the fault.

Examining Faulted Equipment

When a piece of electrical equipment is alleged to have caused a fire, it is important to carefully evaluate the fire scene and remains to determine whether such a conclusion is reasonable. There are a few simple questions that should be asked but often are overlooked:

1. Does the burn pattern at the scene show that the fire was caused by the equipment? The fact that the equipment or an appliance is severely damaged or melted certainly does not prove that it started the fire; such damage is commonly caused by the fire. "V" patterns or low burns on the walls only show that at some time during the fire something burned low in that area or that the flow of hot gases or flame caused such a pattern. In some cases with limited fire damage, the "V" pattern is useful in finding the area of fire origin.
2. How did the equipment cause the fire if it is remote from the area of fire origin? Farfetched theories often are proposed on how device failure in one

part of a room caused a fire in another part of it. The wire insulation usually used in appliances and for branch circuits will melt or burn during a fire but will not support combustion if the heat source is removed. Thus, the burning of wire insulation cannot explain how a fault could create a fire area or origin at a location remote from the faulted equipment. High-resistance faults can cause fires at the location of the fault but not at some remote location.

3. What damage in the equipment indicates that a fault in the equipment started the fire? Energized conductors exposed to heat or flame will arc when the insulation melts; so the mere fact that there are arced conductors present does not mean that the arcing caused the fire. The equipment should be examined for abnormal hot spots or fire damage that might indicate that the cause of the fire was in the equipment. Uniform heating and charring are to be expected after a fire. Evidence of minute arcs or slight abnormal fire damage certainly does not prove that the fire started in the equipment; evidence must exist of sufficient energy in the fault and of combustibles (fuel source) near the fault to justify a conclusion that the equipment started the fire.

4. How did a fault in the equipment cause the fire outside the equipment? If electrical faults are found inside the equipment, then it must be shown how such faults could have caused the fire. There must be combustibles inside or outside the equipment that were ignited. Plastic cases for appliances usually melt readily but will not support combustion. Finding that the plastic housing for an appliance has melted does not prove that the appliance started the fire. Arced wires in the appliance indicate that the power was energized for the appliance but not necessarily that the appliance was the cause of the fire. Conversely, not finding evidence of electrical arcing or electrical faults within the appliance would tend to indicate that the appliance power was turned off when it was engulfed by the fire. Thus, it could not have initiated the fire.

FIRES AND FIRE DAMAGE AS RELATED TO SYSTEM VOLTAGE

The 120/240 V System

Electrical faults in 120/240 V systems often occur at 120 V rather than 240 V because the faults usually involve an ungrounded conductor and the equipment grounding. To have a fault at 240 V requires failure of the insulation on both ungrounded conductors at the same location. A 120 V fault only requires insulation failure at any location on the ungrounded conductor adjacent to uninsulated grounded equipment, such as an equipment chassis, conduit, panelboard, or bare equipment grounding conductor.

It is difficult to maintain an arc in a 120 V system unless there is a fire. As discussed above, two conductors must be in contact with each other to initiate an arc at 120 V. The low impedance of such a fault will produce a large fault current, activate the circuit overcurrent protection device, and extinguish the arc. For a longer air gap, the arc will be extinguished because the voltage is not sufficient to maintain it. In fire conditions, an arc can be maintained for some time when the conductors are not touching. The arc is conducted by the hot ionized gases, and the current is limited by the intrinsic resistance of the arc column.

Higher-Voltage Systems

Higher-voltage systems can produce severe arcing when there is an insulation failure between conductors. Systems above 600 V tend to produce a large-magnitude arcing current that usually activates the circuit protection device. Voltages below 150 V tend to cause arcs that burn out and do not reignite. Between 150 V and 600 V, arcing faults can cause considerable damage, as the arc tends to be maintained with a current magnitude high enough to do considerable damage but not high enough to cause the circuit overcurrent protection device to operate quickly. This phenomenon is recognized in Article 230-95 of the National Electrical Code (1990), which states that ground fault protection of equipment shall be provided for a solidly grounded wye electrical service of more than 150 V to ground but not exceeding 600 V phase-to-phase for each service disconnect rated at 1,000 A or more. Ground fault protection is required because it can be set to operate at lower current values than the normal system overcurrent protection. The reason for these voltage and overcurrent requirements is that arcs at voltages below 150 V will tend to extinguish themselves before extensive arcing damage occurs. Arcs occurring at voltages greater than 600 V will tend to draw excessive current and operate overcurrent protection devices. Arcs in the voltage range of 150 to 600 V will tend to be maintained at current levels of less than 1,000 A and will do considerable damage before circuit overcurrent protection devices can operate. If a ground fault protection device is installed, it can open the circuit before major damage occurs.

A common system voltage used in industrial electrical plants is the 480Y/277 4-wire wye system. When an arc occurs between 277 V and ground, the system is capable of maintaining the arc for a considerable time and can do much damage. Higher system voltages, such as 7,200 volts, for example, will tend to produce large currents and usually will activate circuit overcurrent protection devices in a very short time.

When there is a fire and hot ionized gases are present surrounding energized conductors, arcing can occur between these conductors even though they are not in contact, and no arc would occur across the gap in normal air. An example of such arcing occurred in a power plant where there were 12.47Y/7.2 kV bus bars run along the ceiling in a bus duct. Water dripped on the system and caused an arcing fault at the bus supports, so that an arc was initiated across the support insulation to ground. This led to extensive arcing damage at the bus support; but, in addition, there was arcing between one of the bus bars and the bus duct near the center of the bus run between supports. This arcing at the center of the bus run could not have been the initial fault because there was limited arc damage at this location, and the air gap between the bus and the duct was larger than one through which an arc could be initiated. After the bus failure at the support, however, the smoke and the hot gases in the bus duct were sufficient to allow arcing through the air gap between the bus and the duct.

EXAMINING ELECTRICAL SYSTEMS AND EQUIPMENT AFTER A FIRE

After a fire, the investigator will have to examine the electrical system, the electrical equipment, and the general fire scene. It is important to detect damage or electrical faults that may indicate what could have caused the fire. A thorough examination will require use of the investigator's full technical abilities. The investigation must be supported by knowledge of how electrical equipment and systems operate in normal and fire environments.

Panelboards, Junction Boxes, and Conduit

Following a fire, metal equipment enclosures such as panelboards, junction boxes, and conduit should be examined carefully. It is necessary to determine what internal damage has occurred and how any internal equipment faults could have caused a fire outside the metal equipment enclosure. It is quite common to find holes arced in metal enclosures after a fire because the fire has melted the wire insulation and allowed arcing to occur between energized conductors and the grounded metal enclosure. It is much less common to find such arcing that was initiated prior to a fire, because of the difficulty of initiating and maintaining an arc when a fire is not present.

Inside the enclosure, the equipment must be examined for hot spots or evidence of internal damage that would not be characteristic of an external fire heating up the metal enclosure.

ELECTRICAL EQUIPMENT DAMAGE RELATED TO FIRES

Certain types of electrical equipment usually are severely damaged when exposed to fire even though the damage will not be related to the cause of the fire. Two examples often found in house fires are TV sets and refrigerators. A TV set usually has a plastic enclosure that will melt and be destroyed when exposed to fire even though the TV set is not the cause of the fire. The modern refrigerator has a lighter-gauge steel exterior than that of older-style refrigerators. The interior of modern refrigerators has energy-efficient, plastic insulation rather than the fiberglass insulation previously used. Because of these fabrication changes, a modern refrigerator often will be found severely damaged after a fire. The newer plastic insulations burn more readily than the older-style fiberglass insulation. An investigator must exercise great care in implicating a refrigerator as the origin of a fire just because it is severely damaged.

The severity of damage to any item of electrical equipment after a fire does not necessarily indicate whether or not the equipment caused the fire. Very severe or extensive damage may be related only to the flammability characteristics of the materials used in construction.

EXAMPLES OF 120/240V FIRE INVESTIGATIONS

Microwave Oven

A microwave oven was located in the fire origin area of a restaurant. Though the interior of the oven showed extensive heat and smoke damage, there was no evidence of arcing or electrically damaged components inside the unit. The power cord supplying the oven did show some arcing between conductors, and there was some indication that the oven had been resting on the power cord near the arcing location. The arcing occurred in an area containing combustible paper products and plastic baskets. Thus, the fire probably was caused by an electrical fault in the power cord that ignited the nearby fuel source of paper products and other items.

High Voltage Current-Limited Power Supplies

High voltage current-limited supplies usually are self-protected so that a short circuit at the output will not cause overheating of the power supply and a subsequent fire. For such a power supply to cause a fire, combustible materials (a fuel source) must be present nearby.

A device used to electrocute insects was powered by a high voltage transformer with a 5 kV, 60 Hz open-circuit voltage and a short-circuit current limit of 16 mA. The voltage was applied between two screens. Such a device can cause fires when a large insect or floating combustible debris, such as hay, is ignited between the electrodes and falls to ignite combustibles below the unit.

A neon sign transformer had an output voltage of 12 kV, 60 Hz with the midpoint of the secondary grounded. This provided a voltage to ground of 6 kV. The current was limited to 30 mA. When the neon sign was installed, there was about a 0.5-inch spacing between an exposed electrical terminal and a grounded building structural frame member. Because approximately 75 kV are required to arc 1 inch through air, there would have been no arcing between the sign terminal and the structural member if they had been separated by air. In this case, however, both the sign terminal and the structural member rested on the same wood stud. The current was able to track across the wood stud, ignite the wood, and start a fire.

Television Sets

After a fire, television sets often are found to have extensive heat damage and melted cabinets. Such damage should be expected after a fire. Any investigation of the set should concentrate on locating abnormal hot areas or burning within the set, if any. Uniform heating or charring patterns within the set would tend to indicate that the set did not cause the fire.

Amateur Radio Transmitter–Receiver

An amateur radio transmitter–receiver was found to be badly damaged after a fire. Its metal case was intact, and its interior was uniformly burned. No abnormal hot areas were evident. A small strand of wire, determined not to be part of the set, was stuck to a transformer case inside the unit. It was alleged that the strand of wire had arced to the case of the transformer and caused the fire even though the arcing, if any, was minute. The strand of wire could have fallen into the case through air louvers during or after the fire. It is difficult to see how the fire could have started in this way because the heat produced by the low energy spark or arc was minimal even though the temperature at a point in the spark could be high (2,000–4,000°C). No combustibles were found in the unit, and there was no evidence indicating how a fault inside the unit could have caused a fire outside the case.

Power Cord

The flexible power cord for an electric toaster arced between the two conductors at the strain relief device where the cord entered the toaster. Molten copper was splattered onto the countertop beneath the toaster. The strain relief device had damaged the insulation between the two conductors, allowing their strands to come in contact with each other and arc. The arcing caused the strands to erode so that the arcing current was not continuous once the strands made initial contact. This arcing further damaged the insulation, permitting the arcing to spread to additional strands that also were eroded. The 15 A circuit breaker never tripped because the arcing current was intermittent and noncontinuous, as might occur with solid conductors touching. It is possible for two stranded conductors of an energized power cord to touch each other so that the exposed strands arc and erode away without ever activating a properly installed 15 A overcurrent protection device in the circuit.

Service Drop

The service drop for a 120/240 V service consists of a set of conductors routed overhead from the transformer on the power pole to the service mast of a building. This drop usually consists of two ungrounded insulated conductors and the bare neutral or grounded conductor. There is 240 V between the insulated ungrounded conductors and 120 V between each of the ungrounded conductors and the grounded conductor. The service drop, owned by the electric utility, is connected to the service conductors for the building at the service head on the service mast. The service drop has relatively poor overcurrent protection because the protection is usually provided by a fuse in the primary of the transformer supplying the service drop. This fuse is selected to protect the transformer so that there can be quite large overcurrents in one of the service drops supplied by the transformer before the fuse will open. This problem is recognized in the National Electrical Code (1990) in Articles 230-70(a) and 230-9(a). These articles require that the service disconnect for the service conductor and the overcurrent protection device be located either outside the building or inside the building at the nearest point to where the service conductors enter to provide protection for these conductors inside the building.

Fires have been started when an ungrounded service conductor became loose and its insulation wore away as it contacted and rubbed the aluminum siding or metal flashing on a building. The metal was energized to 120 V, which produced current in the metal because there was a causal connection to ground at the grounded water pipe in one case and the grounded service mast in another. The current flow ignited combustibles adjacent to the poor current path at the service mast and roof flashing connection in one case and at gaps in the metal siding where the current was concentrated in another.

Sometimes during a fire severe arcing is observed along the service drop or service conductors. This can be attributed to the fire having damaged the conductors' insulation, producing the resulting arcing. The poorly protected conductors (from an overcurrent condition) can sustain considerable arc damage during a fire.

REFERENCES

Barnes, C. C. 1966. *Power Cables, Their Design and Installation*. Second Edition. London: Chapman and Hall (Sverak).

Beland, B. 1981. Arcing phenomenon as related to fire investigation. *Fire Technology* 17(3):189–201.

Beland, B 1984. Electrical damages—causes or consequences. *Journal of Forensic Sciences* 29(3):747–761.

Beland, B., C. Roy, and M. Tremblay. 1983. Copper–aluminum interaction in fire environments. *Fire Technology* 19(1):22–30.

Bernstein, T. and P. P. Chang. 1969. Calculation of the self-heating current for a resistance wire temperature sensor. *IEEE Transactions on Industrial Electronics and Control Instrumentation*, IECI-16(1):92–102.

Bernstein, T. 1983. *Electrocution and Fires Involving 120/240 V Appliances*. IEEE Transactions of Industrial Applications, Vol. IA-19, No. 2, pp. 155–159, March/April. New York: Institute of Electrical and Electronic Engineers.

Hilado, C. J. 1982. *Flammability Handbook for Electrical Installation*. Westport, CT: Technomic Publishing Company.

National Electrical Code, 1990, NFPA 70, 1990 Edition. Quincy MA: National Fire Prevention Association.

Sillars, R. W. 1973. *Electrical Insulating Materials and Their Application*. IEEE Monograph Series 14. London: Peter Peregrinus.

Sverak, J. G. 1981. Sizing of ground conductors against fusing. *IEEE Transactions on Power Apparatus and Systems* PAS-100(1):51–59.

8

Lightning Protection for Buildings, Equipment, and Personnel

Theodore Bernstein

Lightning can cause human and animal death and injury and considerable property damage. The investigator needs a knowledge of the electrical characteristics of the lightning stroke as well as the possible stroke (current) paths to ground. Lightning protection can be quite effective when basic physical principles are properly applied. Electrical systems are particularly vulnerable to lightning damage because of the many miles of exposed lines and the potential for surge voltages to be propagated through the distribution system.

LIGHTNING THEORY

The Cloud-to-Ground Lightning Flash

When certain types of clouds and atmospheric conditions are present, it is possible for charges in a thunder cloud to be separated. The mechanism for this process is not fully understood. The lower portion of the cloud usually takes on a negative charge, whereas the upper portion is positive (Williams 1988). This is true in more than 90 percent of the cases, with only a relatively few clouds having a reversed polarity charge separation (Anderson and Eriksson 1980). The charge buildup occurs regardless of what is on the ground when the atmospheric conditions are optimal. In addition to the negative charge on the base of the cloud, there are small pockets of positive charge on it. With a net negative charge on the base of the cloud, there will be a positive charge induced on the earth below; the greater the negative charge, the greater will be the induced positive charge and electric field at the surface of the earth. When the charge separation in the cloud builds up to a sufficient magnitude to cause a large voltage with respect to ground, on the order of 10^9 V, a localized, branched, air breakdown occurs. This breakdown, called a stepped leader, lowers negative charge and extends from the cloud toward the ground in a step of about 50 meters. Then the stepped leader ceases. After a pause of 30 to 50 microseconds, another stepped leader extends from the cloud, which encompasses the original step with all its branches and extends for an additional 50-meter or so step toward the earth.

As the stepping process continues and the stepped leader lowers negative charge as it approaches the earth, the electric field at the earth's surface increases. This produces streamers of positive ions, upward-going leaders, that move a distance from conducting objects on the earth toward the branches of the stepped leader. Objects such as trees, people, rocks, poles, buildings, or a point on the ground itself can have such streamers. Even poor conductors such as a wooden pole, tree, or fiberglass boat can have sufficient surface conductivity when wet to initiate upward-going leaders. The streamers can reach an eventual length of 10 to 20 meters when a branch of the stepped leader approaches (Golde 1975). This stepping process continues until one of the branches of the stepped leader happens to be within 50 to 100 meters of the ground or some object on the ground. At this point, the last step jumps to join a positive streamer from an object on the ground. The bright light of the return stroke then goes up the leader channel as charge is lowered in the leader channel, making all branches bright. The distance for the last jump by the stepped leader from a branch to the object

Table 8.1. Stepped Leader Parameters.

	MINIMUM	REPRESENTATIVE	MAXIMUM
Length of step (m)	3	50	200
Time between steps (μs)	30	50	125
Average velocity of propagation (m/s)	10^5	1.5×10^5	2.6×10^6
Charge deposited in step leader (C)	3	5	20
Current (A)	several hundred		

Table 8.2. Return Stroke Parameters.

	MINIMUM	REPRESENTATIVE	MAXIMUM
Velocity of propagation (m/s)	2×10^7	5×10^7	1.4×10^8
Current rate of rise (kA/μs)	<1	10	<80
Time to peak current (μs)	<1	2	30
Peak current (kA)	—	10–20	270
Time to 1/2 peak current (μs)	10	40	250
Charge transferred (C)	0.2	2.5	20
Channel length (km)	2	5	14

on the ground that is being struck is called the striking distance. The object struck is the one that happens to be closest to a branch of the stepped leader.

The striking distance is related to the magnitude of the peak current in the first return stroke: the greater the current in the first return strike, the greater the striking distance. This return stroke lowers negative charge from the lightning channel and cloud to the earth. The high current in the return stroke is the result of the charge in motion in the lightning channel and results in the bright arc phenomenon called a lightning stroke. The current in this return stroke is unidirectional with a sharp rise to a peak current and then a less rapid decay. Typical times are a rise time to peak current of 2 microseconds and a decay to 50 percent of the peak current of 40 microseconds. Measurements indicate that the median peak current for the first return strike is 30,000 A. The largest value measured is about 270,000 A.

Representative data concerning the stepped leader are presented in Table 8.1, with data on the first return stroke (Uman 1984) provided in Table 8.2. For comparison purposes, the speed of light is approximately 3×10^8 m/s.

Following the first return stroke, the lightning channel luminosity decreases; and there may be no further strokes, or there may be a bright ball of light, called the dart leader, that descends the primary channel, without branching, to the point of contact on the ground. Then the light of a second return stroke ascends this primary channel with great brightness as more charge is lowered. The totality of all the lightning strokes from a region in the cloud to a point struck on the earth is called the lightning flash. Thus a lightning flash consists of the initial

Table 8.3. Dart Leader Parameters.

	MINIMUM	REPRESENTATIVE	MAXIMUM
Velocity of propagation (m/s)	106	2×10^6	2.1×10^7
Charge deposited in dart leader (C)	0.2	1	6

Table 8.4. Lightning Flash Parameters.

	MINIMUM	REPRESENTATIVE	MAXIMUM
Number of strokes per flash	1	3–4	26
Time interval between strokes (ms)	3	40	100
Duration of flash (s)	10^{-2}	0.2	2
Charge transferred (C)	3	25	90

return stroke and all subsequent return strokes, if any. The first return stroke always has the largest peak current magnitude, with a median measured value of 30 kA. Subsequent strokes have a median value of 12 kA (Uman 1987, 124). Data concerning the dart leader are provided in Table 8.3, with data on the lightning flash (Uman 1984) presented in Table 8.4. Not all lightning strokes are between the cloud and the ground; other lightning strokes can be intercloud, intracloud, and cloud-to-space.

Discussing the direction of lightning travel makes little sense unless terms and conditions are very carefully defined. In a usual lightning flash, negative charge travels from the cloud to ground, so the conventional current direction is from ground to cloud. The stepped leader and the dart leader light travel from cloud to ground, whereas the return stroke light travels from ground to cloud. It is probably better to refer to lightning as having a certain path without describing a direction rather than to indicate that lightning traveled in a certain direction along a path between the cloud and the ground.

Attraction of Lightning

Except for the case when tall structures such as radio towers or buildings over 60 meters in height are on the ground under the cloud, the lightning stepped leader will start descending from the charged cloud regardless of what is on the ground. When one of the stepped leader branches happens to be within 50 to 100 meters of the ground or an object on the ground, a return streamer from the object will reach the stepped leader channel and initiate the return stroke. An object on the ground will attract only lightning that would have struck in its immediate vicinity. Thus a sailboat in the middle of a lake will not initiate a lightning strike or attract lightning from all over the lake, but only will attract lightning that would have struck in the vicinity of the sailboat.

For tall structures the stepped leader may start from the structure and travel to the overhead cloud, so that lightning is triggered by the tall structure on the ground. For structures less than 60 meters high, fewer than 10 percent of the flashes are upward (Anderson and Eriksson 1980). Upward flashes have branches extending upward from the tall structure.

Isocenauric Level and Ground Flashes

The isocenauric level for a specific location on the earth is means of indicating the number of "thunderstorm days" per year at that location. A thunderstorm day is defined as a day when at least one clap of thunder can be heard. It makes no difference whether there are many claps of thunder or just one is heard. A lightning stroke produces thunder because of the shock waves created by sudden heating of the air in the lightning path. Peak temperatures have been reported near 30,000°C (Uman and Krider 1989). Because thunder seldom is heard more than about 15 miles away from a stroke, hearing thunder indicates that lightning was within this distance from the observer (Uman 1987). A map showing the isocenauric levels in the United States is presented in Figure 8.1. Isocenauric levels vary from less than five thunderstorms per year along the Pacific Coast to about 100 in Florida.

The isocenauric level is useful in determining regions of the country where lightning may be a problem because of its prevalence. There appears to be a correlation between the isocenauric level and the ground flash density in flashes per square mile or square kilometer per year. The ground flash density is useful

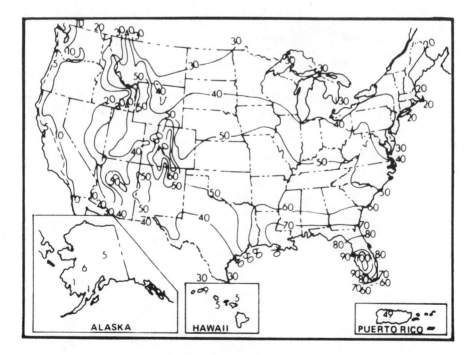

Figure 8.1. Isoceraunic map of the United States.

in evaluating the probability of a lighting strike to a given area on the earth's surface. A method for estimating the ground flash density as related to thunderstorm days is provided in Table 8.5. The data display a wide scatter and may be used only as an approximation (Anderson and Eriksson 1980).

The use of ground flash density to estimate the probability of a lightning stroke in a given area is illustrated by the following example. To determine the number of years between lightning strokes to a given quarter-acre lot, assume that this lot is part of a large region of about the same height. The probability that this lot will be struck by lightning can be evaluated by using the cloud-to-ground lightning flash density in Table 8.5. The data needed for a region where the isocenauric level is 40 are:

Area of lot: 10,890 ft^2
Area of a square mile: 27.878 $\times$ 10^6 ft^2
Isocenauric level: 40
Flashes/mi^2/year: 7.3

Then for this lot,

$$\text{Flashes/year} = (10,890/[27.878 \times 10^6]) \times 7.3$$

$$= 2.85 \times 10^{-3}$$

$$\text{Years/flash} = (2.85 \times 10^{-3})^{-1} = 351$$

From the data it can be concluded that on the average, the lot would be struck by lightning once every 351 years. Because lightning strikes are a random

Table 8.5. Cloud to Ground Lightning Flash Density Related to Thunderstorm Days.

THUNDERSTORM DAYS PER YEAR	GROUND FLASHES/KM^2/YEAR		GROUND FLASHES/MI^2/YEAR	
	NOMINAL	OBSERVED RANGE	NOMINAL	OBSERVED RANGE
5	0.2	0.1–0.5	0.5	0.26–1.3
10	0.5	0.15–1	1.3	0.39–2.6
20	1.1	0.3–3	2.8	0.78–7.8
30	1.9	0.6–5	4.9	1.6–12.9
40	2.8	0.8–8	7.3	2.1–20.7
50	3.7	1.2–10	9.6	3.1–25.9
60	4.7	1.8–12	12.2	4.7–31.1
80	6.9	3–17	17.9	7.8–44
100	9.2	4–20	23.8	10.3–51.8

phenomenon, this lot could be struck two or three time in one year; but, averaged over a time period approaching infinity, there would be about one strike every 351 years. The probability of a strike to a building on this lot is even less than this. It is important to note that this calculation is for a direct lightning strike. A building on the lot could be damaged by a lightning strike through its electrical or telephone system more often than this.

Cone or Zone of Protection

A tall object will tend to attract lightning that would have struck in its vicinity; so for many years various methods were proposed to evaluate the protection accorded to a lower object by a nearby taller object. One suggestion is that there is a cone of protection around an object. In the case of a tall grounded metal pole, an object will not be struck by lightning if it is within the conical volume whose height is the pole and whose base is a circle on the ground, centered on the pole, with a radius equal to the height of the pole (Golde 1975). Lightning would tend to strike the pole rather than an object within this volume. Such a cone of protection is not perfect, and there are numerous reports of objects within the cone of protection being struck.

Using the cone of protection concept, it is evident that the probability of an object being struck is proportional to the square of its height above the surrounding region. A pole of height h will tend to attract lightning to itself that normally would have struck in the circular area at its base that has a radius h. The area is proportional to the square of h.

Objects or people and animals within a zone of protection can be damaged or injured by lightning even if they are not struck directly. This damage is caused by a sideflash or ground currents. When lightning strikes, a large voltage to ground can be developed along the path because of the large peak currents and sharp rate of rise of the current. The peak currents cause high voltages to ground because of resistance in their path such as the resistance of the object struck and the resistance at the ground connection. The rapid rate of rise for the current produces voltages to ground because of inductance in the object struck. When the object that was struck has a high voltage with respect to ground, a sideflash over to a nearby grounded object can occur. The currents in the ground can cause large voltage differences along the earth's surface and can cause damage or injury to objects or people in their vicinity.

Rolling Ball Concept for Protection Zones

Other concepts for zones of protection, taking into account the striking distance for the last jump of the stepped leader, are used in the design of lightning protection systems. A rolling sphere method to determine zones of lightning protection is used in the lightning protection code (*Lightning Protection Code* 1989). This sphere of radius 150 feet (45 m), the striking distance, is imagined as being rolled along the surface of the earth. A zone of protection is provided to the space located under the surface of the sphere when it is tangent to the earth and is resting on a lightning protection air terminal. A zone of protection also is formed when such a sphere is resting on two or more air terminals.

Lightning Nests

It often is suggested that there are certain areas within a region that are lightning nests. In these nests, lightning seems to strike much more often than might be expected. The apparently high lightning strike density cannot be explained by terrain or height factors. Some observers have suggested that these nests are caused by minerals in the ground.

No such things as lightning nests really exist. What appear to be nests are merely the result of the clustering phenomenon noted in the random nature of lightning striking the ground. It is important to note that random does not mean uniform. Random points in a region will tend to cluster and will approach uniformity only when the number of random points becomes large. For example, tossing a coin will produce a heads or a tails result in random fashion. One

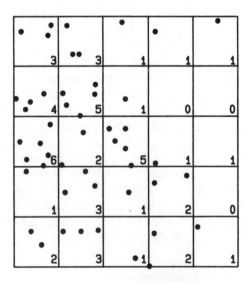

Figure 8.2. Representation of a 10-year period with 50 random lightning strikes to a one-square-mile area. Each small square represents a 0.2 mile X 0.2 mile area. The number in a small square is the number of strikes within the square. If the strikes were uniform, two strikes would appear in each square.

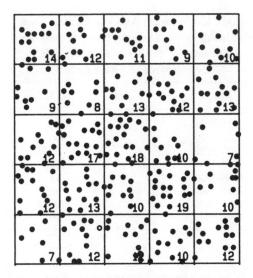

Figure 8.3. Representation of a 60-year period with 300 random lightning strikes to a one-square-mile area as in Figure 8.2. The first ten years are as in Figure 8.2. If the strikes were uniform, 12 strikes would appear in each square.

might get 8 heads and 2 tails in 10 tosses because of the clustering phenomenon. However, in 10,000 tosses the number of heads and tails will be very close to equal.

A grid where 50 points have been selected at random is shown in Figure 8.2. This could represent the lightning strikes to one square mile in 10 years. Note the clustering. People residing in these areas might think they live in a lightning nest. The same grid with 250 additional random points is shown in Figure 8.3, representing the lightning strikes to this one-square-mile area in 60 years. It is evident that the random strikes are now more uniform.

In a region where all structures are of about the same height, lightning strikes are uniform. Any supposed lightning nests are merely the clustering of random strikes (Bernstein 1984).

DEATH AND INJURY CAUSED BY LIGHTNING

The annual number of deaths from lightning in the United States has been trending steadily downward during the twentieth century. Table 8.6 shows that in the years 1910 to 1950 the number of deaths from lightning per year averaged about 300 to 400, but by the 1980s the number had declined to about 90. This is a particularly surprising change considering that it occurred in an ever increasing population. The most likely explanation is that most lightning deaths occur

Table 8.6. Annual Average Deaths from Lightning in the United States. (From *Vital Statistics of the United States*, U.S. Department of Commerce.)

PERIOD	ANNUAL AVERAGE
1910–1919	306
1920–1929	406
1930–1939	400
1940–1949	330
1950–1959	184
1960–1969	133
1970–1979	108
1980–1986	90

outdoors, and early in the century the rural percentage of the population was greater, with more people outdoors and thus more frequently exposed to lightning than is today's mostly urban population.

People Outdoors

People in the open can be killed or injured by a direct stroke of lightning or by ground currents from a nearby stroke. Accidents typically have involved individuals standing outdoors, such as persons at a soccer or a baseball game; people riding on top of hay wagons; mountain climbers caught in storms; golfers; or people closing car windows and standing on the ground when lightning struck the car. The probability that an individual in the open will be struck by lightning is related to the square of his or her height; the person will tend to attract lightning that would have struck the ground within a circular area, centered on that person, with a radius equal to his or her height. The lightning would have struck in that area anyway, regardless of whether the person was there.

A person standing next to a tree has an increased likelihood of being injured by lightning; because the tree is taller than the person, the probability of the tree's being struck is greater than that for an individual standing alone at the same location minus the tree. A tree five times as tall as a person will increase the person's probability of being struck by a factor of 25. When the tree is struck, there can be high voltage-to-ground on the tree trunk, owing to the voltage drop caused by the large peak lightning current in the high-resistance tree and the path from the trunk into the ground. This high voltage can cause sideflash to a person standing next to the tree. In addition, ground currents from the struck tree can produce shocks because of the voltage gradient that may exist along the earth's surface.

Golfers who remain outdoors on a golf course during thunderstorms are at risk. They should leave the course if lightning threatens; the clubhouse is a safe place. Locations that are especially likely to receive a lightning strike should be avoided. Such hazardous locations are found under trees or in small shelter houses without lightning protection. The likelihood of a lightning strike is greater in these places than it is for an individual standing on the gold course, as the height of the tree or the height and the area of the shelter house are greater than the size of the individual.

Telephone-Related Death and Injury

Death or injury from lightning is not common, but people may be killed or injured while using the telephone indoors during a thunderstorm. Each year in the United States one or two people are killed and five are injured while talking on the telephone—a small number, considering the large number of telephones in use (Bernstein 1973).

The telephone lines entering a building must have a protector at the entrance location (*National Electrical Code* 1990). The protector is intended to limit the voltage-to-ground that might occur on the internal telephone wiring or handset because of lightning or power surges; it provides protection if a high voltage power line contacts the telephone line.

A schematic diagram for a typical protector connection is shown in Figure 8.4. The incoming telephone lines are connected at the protector to the telephone grounding electrode by either a gas tube or a carbon block with a spark gap. When the voltage-to-ground on either of the telephone lines exceeds 300 to 400 V, the gas tube fires, or there is an arc at the carbon block spark gap. This tends to hold the voltage across the gas tube or spark gap to the breakdown value and to limit the voltage between the handset or wiring and the connection to the telephone grounding electrode. It is important to note that the voltage-to-ground can be quite high because of the voltage drop caused by the lightning current in the resistance at the telephone grounding electrode and earth interface. For this reason, the National Electrical Code (1990) requires that the telephone ground be bonded to the power ground electrode and the metallic plumbing. This connection serves two useful functions in case of lightning or power surges: the ground resistance will be lowered because of the paralleling of grounding electrodes so that the voltage-to-ground on the telephone wiring will be reduced, and, even more important, the bonding will minimize the voltage difference between the telephone circuit and the power ground or any other grounded objects near the telephone.

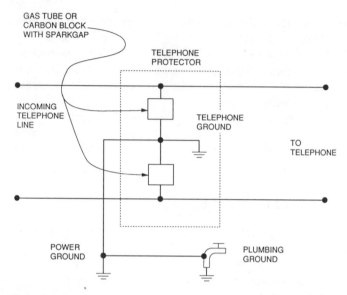

Figure 8.4. Schematic diagram of surge protector for a telephone system.

Bonding of the telephone and power grounding electrodes provides essential protection against death or injury. Consider a situation where the telephone system has a separate driven ground rod that is not bonded to the electrical power grounding electrode system and the metallic plumbing. In such a situation, a voltage surge on either the telephone system or the power system can raise that affected system to a high voltage relative to ground. Because the systems are not bonded together, the unaffected system will remain at substantially zero or ground potential. A person holding the telephone and coming into contact with any grounded object, such as the plumbing, a grounded appliance, or a grounded conductor, can have this large voltage difference placed across his or her body and may be killed or injured.

A voltage surge that occurs when the telephone and the power grounds are bonded together will cause both ground systems to rise together to an elevated high voltage referenced to true ground. However, because of the bond, the voltage difference between the telephone and a grounded appliance or a plumbing fixture will be low, and a shock will not be likely. Most cases of death or injury while a person is talking on the telephone occur because no bonding was installed between the telephone and the power grounds

Important Note. The previous situation and NEC rules apply equally to cable systems, outdoor antenna systems, or any other system required to be grounded. It is especially true for computer and electronic measurement systems. All grounded systems in any building need to be bonded together for safety in order to avoid the voltage difference between different grounding connections. This is a stringent requirement of the NEC.

LIGHTNING PROTECTION FOR STRUCTURES

External Lightning Protection Systems

Lightning protection for buildings is provided by a system of grounded lightning conductors on the structure that provides a low-resistance path to ground for the lightning stroke current. The lightning conductors conduct the lightning current harmlessly to ground and only attract lightning that would have struck the structure anyway. Without lightning conductors, lightning striking a building will find its own path to ground through the poorly conducting structure and will do considerable damage. The current in a lightning stroke is a constant current; that is, the magnitude and the time variation of the current are the same no matter what resistance is in the current path. For this reason, large lightning stroke currents in poorly conducting, high-resistance structures, such as a chimney or a block or brick wall, can cause considerable damage. The power delivered is I^2R; when the current I is very large and the resistance R is relatively large, there will be a large power dissipation that can cause considerable damage. Besides the lightning conductors that prevent damage to the building, a proper lightning protection system also must provide surge protection for the power system on entering the structure.

A modern lightning protection system consists of the following subsystems: air terminals, lightning conductors and down conductors, a grounding electrode, and appropriate bonding.

The air terminals are the lightning rods placed on the structure to intercept the lightning strike. The lightning conductors and down conductors carry the intercepted lightning current from the air terminals to the grounding electrode. The grounding electrode connects the lightning protection system to the earth, and the bonding system makes appropriate bonding connections with the power, telephone, and plumbing grounds. Voltage surge protection prevents voltage surges caused by lightning from being propagated to the internal electrical system. A minimum system consists of surge arrestors placed at the electrical service entrance (*Lightning Protection Code* 1989).

Bonding of the lightning protection system to the other system grounds and bonding of lightning conductors to any nearby grounded objects is essential. Lightning currents in the lightning conductors can produce a large voltage-to-ground on the conductors because of the inductance in the conductors and the resistance in the conductors and the ground electrode ground resistance. The voltage caused by the conductor inductance is proportional to the rate of change of the current, whereas the voltage caused by the resistance is proportional to the instantaneous current magnitude. Large voltages-to-ground on the lightning conductors can lead to sideflash to nearby grounded objects not bonded to the lightning conductor system. If all grounding systems are bonded to the lightning protection system ground, the resistance-to-ground is reduced, and all systems will tend to be at the same voltage-to-ground. This arrangement will reduce the voltage-to-ground on the lightning protection system if there is a lightning strike, and will tend to reduce the possibility of a sideflash to the different ground systems.

Even though lightning stroke currents have a median value of 30 kA, with the largest recorded peak value about 270 kA, an AWG 8 copper conductor with a diameter of 0.129 inch can carry these currents without fusing or melting. This may at first appear strange because an insulated AWG 8 conductor in a raceway with three conductors only has an ampacity of 40 to 55 A, depending on the temperature rating of the insulation (*National Electrical Code* 1990). For a pulse of current with a short duration such as a lightning current, the heating of the conductor is related to the square of the current magnitude and the duration of the pulse. In this case, the current magnitude, on the order of 30 kA, is large, but the pulse duration, about 100 microseconds, is small. Conductors larger than AWG 8 often are used in lightning protection systems for their physical strength.

The lightning protection requirements recommended for the United States are detailed in the Lightning Protection Code (1989). This code was promulgated by the National Fire Protection Association and only has the force of law if adopted by a local governmental jurisdiction. Other countries have developed their own codes. Some standards for lightning protection system components have been developed by Underwriters Laboratories Inc. To improve the quality of the work performed by lightning protection system installers, a certification program was developed by Underwriters Laboratories. In this program, installers must use UL listed components, install their systems according to the Lightning Protection Code, and pass periodic inspections of the quality of their work. A lightning protection system installed under these conditions is called a Master Labeled System.

PROTECTION FROM SURGE VOLTAGES

Voltage Surges Caused by Lighting

Surge voltages in 120/240 V and other low voltage systems result from either lightning or switching transients. Lightning can cause overvoltages by a direct strike to overhead lines or by raising the ground potential of a system as the lightning follows this path to ground. (Lightning and its characteristics were discussed earlier in the chapter.) Lightning voltages caused by a direct strike can be as much as 1 MV or higher if they are not limited by an insulation flashover. Another source of lightning surge voltage is called an induced surge. In an induced surge, lightning does not directly strike the electrical system in question. The voltage surge is caused by the release of the bound charges on the system when the charged overhead clouds and lightning channel are rapidly neutralized by a lighting strike to a point away from the electrical system. Induced lightning surge voltages on overhead lines are on the order of 100 kV (Rusck 1977).

Voltage Surges Caused by Switching Transients

Switching transients can come either from the power system outside the building or from within the building, and can be caused by such sources as the switching of large loads; resonating circuits associated with switching devices; system faults when fuses, circuit breakers, or reclosers operate; or arcing and short-circuit faults. Such surge voltages have lower magnitudes than lightning surges.

Surge Voltage Characteristics in a Building

Surges within buildings rarely will exceed 6 kV on a 120/240 V system. The 6 kV maximum is a result of the usual clearances found in indoor wiring devices, which will tend to sparkover at this level and limit the voltage. In any region with high thunderstorm activity or frequent and severe switching transients, a 6 kV sparkover might be expected to occur about 10 times in a year; but such a voltage peak might occur only once in 100 years in low-exposure areas where there is low thunderstorm activity or minimal load-switching activity (ANSI/IEEE C62.41 1980).

Oscillograph measurements indicate that surge voltages in buildings on low voltage systems have a decaying oscillatory waveform with a frequency of oscillation typically ranging from 30 to 100 kHz. This is true for surge voltages caused by either lightning or switching transients. At the service entrance of a building and outdoors, both decaying oscillatory and unidirectional surge voltages have been observed. Examples of decaying oscillatory and unidirectional waveforms are shown in Figure 8.5. External unidirectional surge voltages due to such source as lightning tend to become oscillatory inside a building because of the natural resonant frequencies of the building's electrical system (Martzloff 1989).

Little attenuation or voltage buildup of the surge voltages occurs within a building when the power lines have only a small load. This difference from usual transmission lines occurs because the lines are short, and the transit times of the surges are short compared to the rise time of the surge. Appliances represent a small load when turned off or when only the control circuitry is on. Some

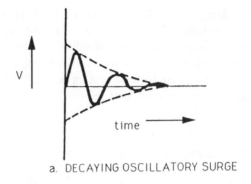

a. DECAYING OSCILLATORY SURGE

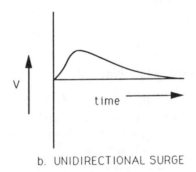

b. UNIDIRECTIONAL SURGE

Figure 8.5. Examples of (a) decaying oscillatory and (b) unidirectional surge waveforms.

attentuation of the surge voltage occurs when the appliance is turned on during the surge.

SURGE VOLTAGE AND CURRENT TEST WAVEFORMS

Recommended test voltage waveforms are useful in determining how well equipment can withstand surges in practice. These test waveforms are representative only of the forms measured in buildings, but they seem to be successful in correlating success in testing with success in application.

For testing major feeders and short branch circuits, such as heavy appliance outlets with short connections to the service entrance, both a unidirectional and a decaying oscillatory waveform are used. Both types of waveforms have been measured in this area. For high-impedance loads, such as those present when an appliance is turned off or when only the control circuit is on, the voltage of the test waveform is specified. For low-impedance loads, the short-circuit current of the test generator also is specified. For the unidirectional test waveform, a

1.2 μs × 50 μs waveform with a 6 kV peak open-circuit voltage for the test generator is recommended for high-impedance loads, where 1.2 microseconds is related to the rise time, and 50 microseconds is the time required to fall to one-half of the peak. For low-impedance loads, the unidirectional test has an 8 microsecond × 20 microsecond waveform with a 3 kA peak short-circuit current for the test generator. The oscillatory test voltage used is a 0.5-microsecond, 100 kHz waveform. This oscillatory waveform rises from 10 to 90 percent of the peak value in 0.5 microsecond, and then decays while oscillating at 100 kHz, each peak being 60 percent of the preceding peak. For the oscillatory waveform and for high-impedance loads the peak open-circuit voltage of the test generator is 6 kV, whereas for low-impedance loads the peak discharge current for the surge is 500 A.

The only test waveform to be used for equipment with a long branch circuit is the 0.5-microsecond, 100 kHz decaying oscillatory waveform. A long branch circuit is defined as a circuit longer than 10 meters with a conductor size between AWG 14 and 10. For high-impedance loads the peak open-circuit voltage of the test generator is 6 kV, and for low-impedance loads the peak discharge current for the surge is 200 A (ANSI/IEEE C62.41 1980).

PROTECTING APPLIANCES FROM LIGHTNING AND SURGE VOLTAGE FAILURES

Protection of an appliance from lightning and surge voltages should be considered during its design—particularly if semiconductor components are involved. The appliance must be able to withstand at least 3 kV voltage spikes and preferably 6 kV voltage spikes for indoor application. Outdoor applications require a 10 kV withstand capability (Martzloff 1982).

Proper bonding and grounding must be used to avoid excessive differences of potential within the appliance. Installers must be educated on proper bonding and grounding during installation. A typical problem occurs when cable TV converters are damaged by lightning because the installer used driven ground rods, not bonded to the power ground, to ground the antenna cable sheath. This can lead to large voltage differences between the cable sheath and the power ground because of surges, with subsequent converter failure. The importance of bonding should be appreciated, as it is not the magnitude of the potentials that is important but the potential difference across a component (Bernstein 1982).

If surge voltage protection is required, it can be one of three types:

1. Filters to attenuate the transient, for example, surge capacitors.
2. "Crowbar" diverters to divert the transient, such as a spark gap, gas tube, or thyristor.
3. Voltage clamping type diverters, such as a zener diode or a varistor.

Surge capacitors commonly are used in low voltage, power circuits. Crowbar diverters are not usually used in low voltage, power circuits because of problems with imprecise triggering and with interrupting follow-current after the diverter has triggered. The crowbar diverter is used for telephone protectors when follow-current is not a problem. Metal oxide varistors recently have found extensive application in voltage clamping.

In an appliance, varistors must be placed across the circuit containing the voltage-sensitive components. In 120 V branch circuits the varistor usually is between the ungrounded conductor and the grounded (white) conductor. It also may be necessary to place varistors between the ungrounded conductor and the grounding (green) conductor as well as between the grounded conductor and grounding conductors.

INVESTIGATING ALLEGED SURGE VOLTAGE FAILURES

Surge Voltage Failure Modes for Appliances

When analyzing the failure of an appliance allegedly damaged by lightning or surge voltages, the investigator must be familiar with the usual failure modes that produce such damage. Lightning can cause failures because of a direct lightning strike or voltage surges created in the electrical system. Direct strikes usually are not a problem for appliances in buildings except for communication equipment without side antennas. For such antennas and the lead-in cable, it is important that the antenna and the cable ground be bonded to the building electrical ground to prevent sideflash. It also is necessary to bond any rooftop appliances to the building ground for the same reason.

Surge voltages will damage appliances by destroying the integrity of the electrical insulation. In solid state control circuitry little visual damage may exist because of the relatively low power level available. Extensive damage can occur in portions of the appliance supplied by the 120/240 V lines because the insulation failure caused by the surge is followed by extensive damage produced by the 120/240 V system fault across the damaged insulation.

Differentiating Surge Voltage Failures from Other Failures

It is difficult at times to determine if an appliance failure was caused by a voltage surge or was just due to overheating, age, moisture, dirt, component wearout, or any of a number of other reasons for normal failure. After the failure, the initiating cause may not be evident.

Surge failure is a likely cause if several electrical appliances are damaged at the same time. Damage to incandescent electrical lamps on the premises is a good indication of over voltage, as they are readily burned out by a high voltage surge. If only one appliance is damaged in a building, one must ask why the alleged surge voltage was so selective. If there is no good answer to this question, then normal failure might be assumed. This is particularly true for an appliance that was near the end of its normal life.

Open windings or conductors probably are not caused by the voltage surge itself, as the time of such a surge is too short for sufficient heating of the conductor to cause it to part. Some insurance claims for lightning damage to hermetically sealed compressors have been made for units that showed no external damage. Internally, the bearings had seized, or a motor winding was open, and such damage probably is not due to lightning.

SOME TYPICAL MISCONCEPTIONS ABOUT LIGHTNING

1. *People are safe in an all-metal car because the rubber tires insulate the car from ground.*

People are safe in an all-metal car when lightning strikes the car because the car's metallic body will act as an equipotential surface. The voltage difference between any two points of contact with the car body will be very small. There will be a current path through the metal body of the car and over or through the tires to ground. Usually, at least one or more of the tires will be blown or severely damaged by the high voltage and current associated with the lightning strike. The metal body of the car will have a high voltage relative to ground because of the large resistance at the tire–ground interface. However, the voltage difference between any two points on the metal body of the car that an occupant might contact would be small. However, there is danger for a person standing on the ground and touching a car when lightning strikes. In this case, the voltage between the person's hands on the car and feet on the ground can be quite high and may lead to electrocution.

2. *A golfer was struck by lightning because he was wearing shoes with cleats or was carrying a metal golf club.*

What a golfer wears or carries has little effect on the probability of his or her being struck by lightning. There is enough conductivity in a person's body to initiate an upward-going leader. Holding a metal golf club high above one's head would increase the probability of being struck. Whether a golfer was wearing regular shoes with cleats or even shoes with rubber soles would make no difference, as the ground resistance at the golfer's feet after the strike would be about the same. The rubber soles would not serve as effective insulation to prevent a lightning strike, as there would be enough conductivity in the golfer's body and over the rubber soles to initiate a leader and attract a lightning stroke. Recall that lightning has traveled miles in space, so there would be no problem in its piercing some rubber soles. The cleats on golf shoes have a negligible effect on the overall ground–feet interface resistance.

3. *Lightning conductors attract lightning to a building; so it is better not to have lightning conductors.*

Lightning conductors attract lightning that was going to strike the building or in its immediate vicinity anyway. It is far better to provide a safe, low-resistance path to ground for lightning to take rather than to have it select its own path through the interior of the building to ground. The latter course can be very destructive.

4. *A farmer was pleased that storm clouds usually came from the west. To the west of the farm home were power lines, which he felt would attract the lightning first and clean out the lightning from the clouds so that his house would be safe from a lightning strike.*

The power lines only would attract lightning that was going to strike in the immediate vicinity of the lines anyway. Unless the farmer lived in the shadow of the lines, they would have no effect on the probability of a lightning strike to his house.

5. *A woman was concerned that she lived near a tall TV tower. She felt that the tower would attract lightning to her home.*

The tower might attract lightning to itself or even initiate a stroke to itself, but it would not increase the probability of the home being struck. If anything, the tower might decrease the probability that lightning would strike the home, by acting like a lightning rod protection system.

6. *Lightning causes power lines to fall because the arc produced by the lightning melts the conductors and causes them to part.*

When lightning strikes a power line, it can cause an arc between conductors or from a conductor to a grounded tower. The usually severe damage to the conductor or tower is caused by the 60 Hz power current, which follows in the arc track initiated by the dielectric breakdown of the air from the lightning strike.

REFERENCES

Anderson, R. B. and A. J. Eriksson. 1980. Lightning parameters for engineering applications. *Electra* 68: 65–102.

ANSI/IEEE C62.41-1980. 1980. *IEEE Guide for Surge Voltages in Low-Voltage AC Power Circuits.* New York: Institute of Electrical and Electronics Engineers.

Bernstein, T. 1973. Effects of electricity and lightning on man and animals. *Journal of Forensic Sciences* 18(1): 3–11.

Bernstein, T. 1982. Proper grounding of cable TV systems. *Communication Engineering Digest* 8(6): 41–47.

Bernstein, T. 1984. Lightning and power surge damage to appliances. *IEEE Transactions on Industry Applications* IA-20(6): 1507–1512.

Golde, R. H. 1973. *Lightning Protection*. London: Edward Arnold.

Lightning Protection Code 1989. NFPA 78 1989 Edition. Quincy, MA: National Fire Protection Association.

Martzloff, R. D. 1982. The propagation and attenuation of surge voltages and surge currents in low-voltage ac circuits. Paper read at the IEEE Power Summer Meeting. Paper 82SM 453-9.

Martzloff, R. D. 1989. The development of a guide on surge voltages in low-voltage ac power circuits. In *Transient Voltage Suppression,* Fifth Edition. Melbourne, FL: Harris Semiconductor.

National Electrical Code 1990. NFPA 1990 Edition. Quincy, MA: National Fire Protection Association.

Rusck, S. 1977. Protection of distribution lines. In *Lightning,* Vol. 2, ed. R. H. Golde, pp 747–771. New York: Academic Press.

Uman, Martin A. 1984. *Lightning*. New York: Dover.

Uman, Martin A. 1986. *All about Lightning*. New York: Dover.

Uman, Martin A. 1987. *The Lightning Discharge*. New York: Academic Press.

Uman, Martin A. and E. Philip Krider. 1989. Natural and artificially initiated lightning. *Science* 246(4929): 457–464.

Williams, Earle R. 1988. The electrification of thunderstorms. *Scientific American* 259(5): 88–99.

9

Static Electricity: Causes, Analysis and Prevention

Glenn Schmieg

FUNDAMENTALS OF STATIC ELECTRICITY

Static electricity is extremely common in the home, industrial settings, and nature. It exists in small amounts when people comb their hair, in larger amounts in sparks from machinery, and in gigantic quantities in the form of lightning. Static electricity is harmless in small amounts, but in moderate amounts and in the presence of flammable materials it can start fires (see NFPA publications in the chapter-end references).

A few basic concepts and definitions must be presented to explain how static electrical charges develop and accumulate and the conditions that can render them harmless. These concepts are charge, electrical forces, conductors, and insulators. (An additional discussion of basic concepts is presented in Chapter 1.)

Charge

Electrical charge is the fundamental quantity of electricity. It is measured in units called coulombs, abbreviated C. Charge occurs in two forms, positive and negative. Both are common, and both can create sparks and cause fires. Negative charge is present in all atoms (the electrons); positive charge is present in the nuclear core of every atom (the protons).

Although charge is present in all matter, most common objects are electrically neutral. The positive and the negative charges that are present exist in equal (and opposite) amounts. Birthday cakes, beer cans, and balloons all ordinarily are uncharged, as is the earth. However, it is easy to disturb this neutral arrangement. When two different items are brought together and then are physically separated, there often is an accompanying separation of their electrical charge. That is, one object may end up with a net positive charge and the other with a net negative value.

You probably have experienced such charging by rubbing a balloon on your hair. The balloon and your hair each become charged. What is the sign of these charges? Does the balloon become negative, or does the hair become negative? Many measurements have been made on human hair, rubber, and other materials in an attempt to answer such questions. Some of these measurements are summarized in Table 9.1.

Table 9.1 lists a variety of materials arranged in relative order of their charging potentials, an order known as a triboelectric series. If any two materials listed in the table are brought into contact or rubbed together, the material higher in the table will take on a positive charge, and the other material (lower in the chart) will become negatively charged. These tables have been known for over two centuries, and one often sees entries that hint of their long history. Silk, cat

Table 9.1. Triboelectric Series.

More Positive
Rabbit fur
Glass
Nylon
Wool
Cat fur
Cotton
Silk
Lucite
Dacron
PVC
Polyethylene
Rubber balloon
Hard rubber
Teflon
Saran wrap
More Negative

fur, and hard rubber were used by early experimenters to produce reasonably reliable "frictional electricity."

Today the value of a triboelectric series is increased if it includes plastics and other modern materials that an experimenter is likely to use. However, even with these newer materials, a note of caution is in order. The triboelectric series was developed by using quite pure, clean materials. If Teflon and wool are to behave as the series predicts, one must use clean Teflon and clean wool. Sometimes the presence of grease, a thin film of moisture, or other surface contaminants can change or even reverse the expected results. Also, the relative humidity can affect results.

Example. A new sorting table is being designed for use in a manufacturing operation. Several cams and wheels will be used that slide over each other. What design features might be used to help reduce the buildup of static electricity?

Solution. If materials are chosen for the cam and wheels that are close to each other in the triboelectric series, the tendency for static charge generation will be minimized. If static generation still is a problem, then lubrication may solve it. Remember that lubrication can keep the parts separated, and the dissimilar materials will not make contact with each other.

Electrical Forces

Electrical charges exert forces on each other. The forces are attractive when one charge is positive and the other is negative; a pair of opposite charges will attract each other and move closer together unless held apart by something else. Repulsive forces also exist, in two ways: two positive charges repel each other, and two negative charges repel each other. Thus, like electrical charges repel each other, and opposites attract.

One cannot see electrical charges directly, but their effects are obvious. There are many common experiments, such as picking up small bits of paper with a charged comb, that demonstrate the existence of these electrical forces. In a home or factory, the presence of static electricity often is deduced from the effects of these forces. One might observe the "static cling" of a garment, feel the hair bristle on one's arm, or see the motion of small pieces of lint on an assembly line.

Conductors and Insulators

A material that allows charge to move freely through it or over its surface is called an electrical conductor. All metals including copper, steel, aluminum, and tin are good electrical conductors. Charges can be generated in one place and then carried along a conductor to produce an effect at another location. The human body is a good conductor, as anyone has learned who has felt an electrical shock. Charge can be transferred from one person to another by a simple touch or a handshake.

Water also is a good conductor. With respect to static electricity, water easily carries charge from one point to another. Sometimes it is said that pure water does not conduct electricity, but this statement is not true in the present context. River water is less conductive than seawater, tap water is less conductive than river water, and distilled water is less conductive than tap water; but all these types of water will easily transfer charge.

Quite a few common materials are insulators or nonconductors. Examples include rubber, glass, wax, paper, and most plastics. These materials do not allow free movement of charge through them or over their surfaces. Any charge deposited on an insulator will remain there for some time. A charge sometimes will stay on a glass surface for hours, and some plastics will store a charge for days, if kept dry.

Because it is a conductor, water can change the nature of other materials. For example, dry cardboard is an insulator, but if slightly moistened, it will readily conduct electricity. Consider the following possible chain of events. Cardboard is present in a room with normal moisture and humidity and is moderately conductive. A fire starts, generating heat that dries the cardboard—which becomes an insulator. A firefighter sprays the cardboard with a fire hose, and it becomes a conductor again. Hours or days later, it may dry out again from normal evaporation and may end up an insulator. The moral of this story is clear:

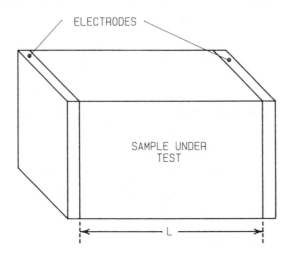

Figure 9.1. Resistivity measurement.

Table 9.2. Electrical Quantities.

QUANTITY	METRIC UNITS
Resistance	ohms
Conductance	siemens
Resistivity	ohmmeters
Conductivity	siemens/meter

in any investigation of a fire, an explosion, or a hazardous situation, one must carefully consider the entire history of all the materials that may have been involved with static electricity.

To test the conduction of a material sample, electrodes are placed at two points, as shown in Figure 9.1. A small test current then provides a measure of either resistance in ohms or conductance in siemens. These two measures are reciprocally related: resistance equals unity divided by the conductance.

An alternate approach defines a resistivity, ρ, that is related to resistance by the formula:

$$\text{resistance} = \rho \, L/A$$

where L is the length and A is the cross-sectional area of the sample under test. The units of resistivity are ohmmeters. Conductivity is defined as the reciprocal of resistivity:

$$\text{conductivity} = \frac{1}{\text{resistivity}}$$

and the units are siemens/meter (s/m).

The four quantities introduced above are listed with their metric units in Table 9.2.

THE IGNITION MECHANISM

Before a fire or an explosion can be caused by static electricity, four conditions must be satisfied simultaneously. They are:

1. A method of static charge generation.
2. The ability to store charge at a voltage.

3. A minimum amount of stored energy (ignition energy).
4. Creation of a spark in a flammable environment.

Each of these four conditions will be considered individually.

Charge Generation

The natural state of matter is one of electrical neutrality. Every neutral atom contains positive and negative charges in equal amounts. Charging an object means effecting a separation of its charges. One body or portion of a body can be made positive, while another is made negative. This charge separation occurs by touching or bringing materials into contact with each other and then separating them again.

The strength of this charging is extremely variable. With some metal surfaces it may be so small as to be difficult to measure. However, if one rapidly pulls two strips of sticky tape apart, the charging may be strong enough to produce tiny visible sparks in a dark room!

The possibility of static electricity generation should be considered whenever materials are brought together and separated again. Common examples are conveyor belts in motion, powdered materials blown through a pipe, and pressurized steam escaping from a leaky vessel.

Most accidental or unintentional charging is done by contact followed by separation. This method is generally unreliable, and is especially so under variable conditions. For example, consider what happens when a person slides across a car seat. On some days, considerable charge is generated. As one slides to a new position or steps from the car, a small but quite noticeable spark jumps from one's hand to a part of the car. It might even be strong enough to make one flinch or exclaim out loud. On other days, the same motion produces no significant charge or spark. As has been noted, a small amount of surface contaminant can change the charging ability by a large amount. Therefore, when one wishes to generate charge deliberately, a more reliable method of charge generation is used.

In the induction method, one charge is used to produce charges at other locations. As shown in Figure 9.2, if a charge is brought near the left end of a neutral conducting rod, the charges internal to the rod will experience a separation

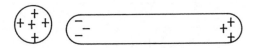

Figure 9.2. Charge induction, Example 1.

toward the two ends. This separation of charges is caused by the electrostatic forces between the charges. The separation will last as long as the external charge remains near the rod. It is said that the rod has induced charges on it.

Notice that the net charge on the rod has not changed. The positive and the negative charges have been induced to separate within the rod, some moving to the left and some moving to the right. When the external charge is taken away from the vicinity of the rod, the separated charges will move back toward each other, recombine, and leave the rod electrically neutral again.

Often, induced charge formation is temporary. For example, a slightly charged cloud passing overhead induces charge formations in the earth below; after the cloud passes, the induced charge formation disburses. Another example is that of an individual shuffling along a linoleum floor; each conductor that is passed reflects the person's movement, producing its own induced charge formation. These are well-defined charges, but they are only temporary.

A trick can be used to create some permanently induced charges. Consider Figure 9.3. It has two external charges, with two rods and four induced charges. So far, this resembles Figure 9.2; if the two external charges are removed, the two rods will be left as they began. Now the trick: While the external charges are present, a short piece of wire is connected between the nearest ends of the rods, and this wire allows the positive and negative charges near the center of Figure 9.3 to combine, or to cancel each other. Then when the wire is removed, each of the two rods has acquired a charge, and these induced charges are quite permanent. When the external charges are removed, the left rod retains a net negative charge, while the right rod retains a net positive charge.

Although Figures 9.2 and 9.3 are abstract representations, they can be used to represent and study many situations found in real life. Moving people and moving machinery parts produce induction charging, and by repetitive induction charging, large amounts of charge can build up. A Dirod generator, shown in Figure 9.4, can be turned by hand or connected to a small motor to provide large amounts of static electricity on demand for experiments or tests. This generator separates charges by repetitive induction. Each small rod repeats the process, as discussed above. Static charges may build up on such a generator to a maximum of 40,000 or 50,000 V. This maximum electrical pressure is easy to measure and describe for most generators of static electricity.

There are at least two ways to express the quantity of charge generation. First, one could measure the time rate at which charges are being separated or

Figure 9.3. Charge induction, Example 2.

Figure 9.4. Dirod generator. (Photo by G. Schmieg.)

collected by the generator. This could be stated in coulombs per second to emphasize the role of charge; or, because one coulomb per second is the same as one ampere of current, it also could be stated in amperes to emphasize that it is a current. All generators of charge have a current associated with them. However, the current may be quite variable or difficult to measure.

A second quantitative measure of charge generation would be the maximum voltage or electrical pressure that could be developed. For example, a certain pair of shoes sliding along a carpet might be able to create a maximum of 6000 V. A higher voltage might create a small spark discharge, followed by another voltage buildup to 6000 volts, and so on.

Charge Storage

When charge generation occurs, the charge may be immediately dissipated, or it may be stored. Storage can occur on an insulator (where the charge is not free to move away) or on an insulated conductor. As charge is added, the voltage or electrical pressure builds up. The charge and the voltage are directly proportional. The amount of stored charge, Q, and the voltage, V, are related by a formula:

$$Q = CV$$

where C is the capacitance, measured in units called farads. The magnitude of the capacitance depends on the geometry of the parts (the size and the shape).

The formula uses Q in coulombs, V in volts, and C in farads. For many engineering situations, approximate values may be used.

Example. If a charge of 3.5×10^{-9} C is placed on a one-inch-diameter steel ball bearing (capacitance $= 1.4 \times 10^{-12}$ F), what voltage results?

Solution:

$$Q = CV$$

$$(3.5 \times 10^{-9}) = (1.4 \times 10^{-12}) V$$

$$V = 2500 \text{ volts}$$

In using the formula $Q = CV$, we can easily calculate any one of the variables if the other two are known. But, which variable usually is known, or can be calculated or estimated? In the example above, the capacitance served as a starting point. A standard formula in electrical theory states that the capacitance of a sphere of radius R is given by:

$$C = 4\pi\varepsilon_0 R$$

where the capacitance C is in farads, the radius R is in meters, and $\varepsilon_0 = 8.8 \times 10^{-12}$. Thus, for a radius of one-half inch (0.013 meter):

$$C = (4)(3.14)(8.8 \times 10^{-12})(.013)$$

$$C = 1.4 \times 10^{-12} \text{ F} = 1.4 \text{ picofarad}$$

For a sphere that is 10 times larger, the capacitance is also 10 times greater, and similarly for any size.

Although this formula is only *exactly* true for an isolated perfect sphere, it can be used to estimate reasonable values for many common objects. Table 9.3 displays a few representative values for capacitance. Because the capacitance values of Table 9.3 are not exact, they should be used in a conservative manner whenever safety is a concern.

Table 9.3. Capacitance Values for Common Items.

ITEM	CAPACITANCE (pF)
Small hand tool	10–20
12-quart pail	30–50
55-gallon drum	50–100
Human body	100–300

Ignition Energy

If charge continues to be added to a body, the voltage will continue to rise, and eventually a spark may be released. To cause ignition, this spark must possess a minimum energy, the exact value of which depends on what materials are being ignited. The energy E released in a spark discharge is given by:

$$E = \tfrac{1}{2} CV^2$$

If C is in farads and V in volts, the energy E will be in joules. Common submultiples include mJ (millijoules) and μJ (microjoules).

Example. Find the energy stored on the charged ball bearing in the example above where $V = 2500$ volts and $C = 1.4 \times 10^{-12}$ F.

Solution:

$$E = \tfrac{1}{2} CV^2$$

$$E = (0.5)(1.4 \times 10^{-12})(2,500)^2$$

$$E = 4.4 \times 10^{-6} \text{ J}$$

$$E = 4.4 \text{ μJ}$$

The formula $E = \tfrac{1}{2}CV^2$, used above to calculate energy storage on a charged ball, applies to the energy stored in a capacitor. When the formula is used as is, the implicit assumption is that all of the stored electrical energy is converted into the energy of the resulting spark. This assumption frequently has been questioned, for quite valid reasons. First, the wires or conductors that make up the discharge circuit will limit and absorb some of the energy because of their intrinsic resistance. This absorbed energy (which appears as heat in the wires) is not available to the spark. Second, experiments with different electrode materials and different electrode spacings have shown that there is some dependence on these other variables. Therefore, the formula should not be used for very low voltages and small spacings. One recommended practice for static electricity, as delineated in the National Fire Protection Association publication NFPA 77 (1988), suggests "that sparks arising from potential differences of less than 1500 volts are unlikely to be hazardous in saturated hydrocarbon gases because of the short gap and heat loss to the terminals."

Example. It is known that air mixed with saturated hydrocarbon gases can be ignited with sparks of about 25 millijoules of energy. What level of voltage on a human body would store this amount of energy?

Solution. Using an approximate value of 300 picofarads (pF) for the capacitance of a human, one finds that:

$$E = \tfrac{1}{2} CV^2$$

$$0.25 \text{ mJ} = \tfrac{1}{2} (300 \text{ pF}) V^2$$

$$(0.25 \times 10^{-3}) = \tfrac{1}{2} (300 \times 10^{-12}) V^2$$

$$V^2 = 1.67 \times 10^6$$

$$V = 1{,}290 \text{ volts}$$

Because this voltage is less than 1,500, it may not cause ignition. However, it is important to note that using a lower estimate for the capacitance of the human body would produce a higher estimate for the voltage.

The energy stored by a capacitor is stored in the electric field produced by the capacitor. An analogous situation exists for an inductor that stores energy in its magnetic field. The amount of energy stored in such a magnetic field is given by:

$$E = \tfrac{1}{2} LI^2$$

Although in principle this stored energy can produce a spark and cause a fire or explosion, it will not be explored here. Currents of this type usually are considered outside the domain of static electricity.

Flammable Mixtures

Many gas mixtures, such as acetylene–air and hydrogen–oxygen, ignite readily. Similarly, dust layers on a surface and dust clouds dispersed in air can be ignited. Experimental values for the minimum energy required to cause ignition are known for many industrial chemicals. Most often these values have been determined by using a small test chamber in which the relative concentration of the gases can be varied. In a few special cases, the oxidizing gas might be varied, or pure oxygen could be used. Generally, however, flammable vapors or dust clouds mixed with air are tested.

Values quoted in the research literature have been compared and criticized many times. For the most part, published tables show minimum energies only, even when it is known that the electrode material, the spacing, and the size affect the results. Therefore, in practical considerations involving safety, consideration should be given to the actual existing conditions. Table 9.4 for vapors and Table 9.5 for dust clouds may be used as guides to supply approximate values. More complete tables, with additional references, are given by Cross (1987), Haase (1977), and Nagy et al. (1968).

Table 9.4. Ignition Energies
for Vapors.

VAPORIZED MATERIAL	MINIMUM IGNITION ENERGY (mJ)
Aviation gasoline	0.2
Ethylene	0.07–0.08
Hydrogen	0.011–0.017
Methane	0.28–0.39
Propane	0.16–0.25

Example. It was shown previously that a one-inch-diameter ball bearing, when charged to 2500 V, stores an energy of 4.4 μJ. Would this situation cause ignition of a mixture of hydrogen and air?

Solution. The minimum ignition energy for hydrogen–air mixtures has been experimentally determined to be in the range of 0.011 to 0.017 mJ (or 11 to 17 μJ), as is shown in Table 9.4. Thus, if charged to 2,500 V, the ball bearing would not cause ignition. However, consideration should always be given to a margin of safety. As the consequences of a fire or an explosion increase, so should the margin of safety.

Example. When one removes one's shirt, considerable charge separation can be produced. Several investigators have found the resulting voltages on experimental subjects to be as great as 10,000 V. Would this voltage be sufficient to ignite typical flammable gases?

Table 9.5. Ignition Energies for
Dust Clouds.

DUST CLOUD MATERIAL	MINIMUM IGNITION ENERGY (mJ)
Aluminum	50–280
Black powder	320
Bone glue	140
Chocolate crumbs	100
Cork dust	35–45
Soap powder	60–960
Wood dust	20–40

Solution. Using a capacitance of 300 pF for the human body yields:

$$E = \tfrac{1}{2} CV^2$$
$$E = \tfrac{1}{2} (300 \times 10^{-12}) (10^4)^2$$
$$E = 15 \times 10^{-3} \text{ J}$$
$$E = 15 \text{ mJ}$$

This energy, 15 mJ, is well beyond the minimum ignition energy for most common flammable gases. However, Table 9.5 shows that this value, 15 mJ, is below that needed to ignite ordinary dust clouds. However, in critical situations one should always be conservative and allow for a margin of safety.

DISSIPATION OF STATIC CHARGES

There are ways to eliminate or mitigate the effects of undesirable or dangerous static charge buildup. The methods that have been studied and developed offer considerable variety. If one seems impractical for a particular situation, another one should be considered. Occasionally, two or more methods of mitigation can be used simultaneously with good effect.

A good safety rule (and its corollary) needs to be established before details of the methods for mitigating the effects of static electricity are discussed: *When one cannot absolutely eliminate a dangerous situation, one must make the occurrence of that situation a very unlikely event.* Because static electricity can seldom, if ever, be completely eliminated, one often must deal with other variables as well. For example, when faced with the possibility of a spark in a flammable atmosphere, one first should work to make the spark unlikely. Then, efforts should be made to make the flammable atmosphere unlikely. For this purpose, ventilation or the addition of inert gases should be considered. The rule's corollary is this: *If two independent events are each unlikely, the combination of both events is very unlikely.*

Humidity

If an area's relative humidity is kept at a high level, some materials will adsorb enough water vapor from the air to become slightly conductive. This moisture may be enough to carry away static charges as fast as they are generated. Paper, cardboard, and glass all adsorb moisture readily. At a relative humidity of 65 to 70 percent, a static charge buildup on these materials usually is small or nonexistent. Even at a relative humidity of 50 percent, there often is significant attenuation of the static electricity.

There are some materials, such as Teflon and other plastics, that do not adsorb water vapor from the air. These materials make good insulators even under conditions of extreme humidity (80–95 percent). To discharge Teflon, an alternate method should be chosen from those that follow.

For extensive or long-term continuous control of humidity, a central furnace or air conditioning system unit may be required. However, in smaller operations a few wet rags or a pan of boiling water may be enough to give relief from electrostatic problems.

Ion Generation

The addition of ions to the air causes it to become conductive. Ions can be readily produced by high voltage sources, flames, and some radioactive sources. A sufficient ion concentration in the air will allow discharge currents to flow through the air and prevent static charge from accumulating.

A commercial ion generator that is used for static control is shown in Figure 9.5. The rectangular box is a high voltage transformer, which is connected to a series of sharp points along a wand. The high voltage (AC) produces ions that stream off into the air from the end of each point. Because both positive and negative ions are created, there is some recombination. However, this effect

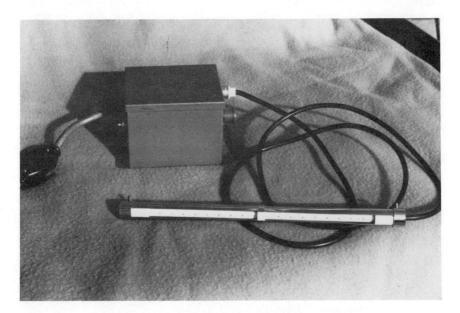

Figure 9.5. Commercial ion generator. (Photo by G. Schmieg.)

usually is of little consequence. The effective range is seldom more than a foot or two, so the ionization units often are located within a few inches of the charged material.

If a region of unwanted positive charge is exposed to an ion stream, the negative ions will be attracted, and neutralization will result. Then the positive ions will be repelled; once neutralization has been achieved, there is no more attraction. Thus, there is a built-in self-limiting feedback mechanism.

Several types of specialty ionizers exist. One type has an air-assist from a small blower to carry the ions a greater distance. It might be recommended near fast-moving or dangerous machinery. Another type has positive and negative (DC) generation from separate power supplies. Each can be adjusted to achieve an exact balance between the positive and the negative ions, or a deliberate bias can be chosen. This action might be recommended when a troublesome process creates a majority of one kind of charge.

If a grounded conductor is brought near an ionizer, a corona current or spark may occur. For this reason, regular ionizers should not be used in flammable atmospheres. However, specially designed ionizers can be purchased that are intrinsically safe. In these ionizers, the energy available for a corona or a spark is insufficient to set off flammable mixtures.

Direct Control of Conductivity

Sometimes direct chemical changes can be made if they do not interfere with the other properties of the process. For example, cloth made of nylon, rayon, and polyester charges easily, but this tendency to charge can be reduced by making the material more conductive. Metal fibers have been tried as additives woven right into the cloth. The cloth still feels good and is washable, but the charging is greatly reduced.

Another approach, which is used with some plastics, is the addition of carbon to increase conductivity. When this is done, the material's resistivity can be varied over an extremely wide range. The lower the resistivity is, the less the charging.

A strip of conducting plastic is shown in Figure 9.6. Note the alligator clips. The conducting plastic strip functions much like a copper wire. It is capable of draining away unwanted charges.

Sprays and dips may be used for surface coatings. In some cases, these added materials are themselves conducting. Water-absorbing materials also are used. For the latter, enough humidity must be available to supply the water, or the spray will be ineffective. Surface coatings must be renewed periodically to accommodate wear. Some spray manufacturers recommend respraying carpets every two to four weeks, depending on their wear.

Figure 9.6. Plastic conducting strip. (Photo by G. Schmieg.)

Grounding

If charges build up on isolated conductors, the conductors should be grounded. By connecting these conductors to ground, the charges are carried away to the earth. Often this grounding can be achieved with a connection to the nearest water pipe. Even a relatively high-resistance path to ground is adequate to bleed away most static charges. One million ohms or less usually is effective.

If charges are accumulating on an insulator, grounding of the insulator will not cure the situation. One of the other methods presented above should be used.

Occasionally it is desirable to ground personnel. This is especially useful around sensitive electronic apparatus or around manufacturing operations with solid state devices. A note of caution for such cases: Do not attach very low-resistance grounds to people. There is always danger of an electrical shock by accidental contact with a source of 110 V AC. There must be no risk of a high current flow through the body in these cases. Make sure that any ground connection involving people has a resistance greater than 25,000 ohms.

Passive Discharge

Any sharp conducting point will emit a corona or send out ions if the voltage is raised beyond a few thousand volts. For this reason, points or wire brushes sometimes are fastened to equipment to prevent static buildup. As the voltage rises, the points begin discharging. One advantage of this arrangement is that no external power source is needed. One disadvantage is that complete discharge never is achieved. The system may remain at a potential that is sufficient to cause problems. Another disadvantage is the danger that the sharp points present to work personnel.

INSTRUMENTS FOR MEASUREMENT OF STATIC ELECTRICITY

Many commercial meters are available for measuring charge, current, voltage, and other electrical parameters. In general, the cost increases with the accuracy and the sophistication required of the measurement. For example, a small hand-held meter with self-contained battery costs about $30 and measures resistance, voltage, and current in several ranges. However, such an inexpensive instrument is useless for most resistance measurements of static electricity. The internal 1.5 V battery is used to set up small test currents for measuring resistance, and this small voltage can easily be "fooled" by a small piece of dirt or surface contamination. Therefore, resistance measurements for electrostatics should be made with a test voltage of at least 500 V. The higher voltage can break through the small surface imperfections and provide more accurate information than the small voltage can give.

Voltage Measurements

Electrostatic voltmeters provide a noncontacting means of measuring the electrostatic surface potential of insulating, conducting, or semiconducting materials. These devices are distinguished by their high accuracy (0.1 percent or better) and their ability to resolve targets as small as one millimeter in diameter. Typical applications include measurements on electrophotographic materials and fabrics, as well as the monitoring of charge accumulation.

Field Measurements

A voltmeter measures the actual potential at the surface of the material under test. A fieldmeter measures the electrostatic field (in volts per meter) at the location of a special grounded probe. The probe-to-surface separation must be carefully controlled for accurate measurement. Electrostatic field strengths of several hundred kilovolts per meter can be measured with large separations. Applications include safety monitoring in explosive gas environments and atmospheric electricity measurements.

In Figure 9.7 a portable electrostatic fieldmeter with several available ranges on logarithmic or linear scales is shown. Photograph provided through the cour-

Figure 9.7. Portable electrostatic fieldmeter.

tesy of Monroe Electronics. Less than one microjoule of spark energy is available at the probe.

REFERENCES

Cross, J. A. 1987. *Electrostatics: Principles, Problems, and Applications.* Bristol U.K.: Adam Hilger.

Fire and Dust Explosions in Facilities Manufacturing and Handling Starch. NFPA 61A 1989 Edition. Quincy, MA: National Fire Protection Association.

Fire and Explosions in Feed Mills. NFPA 61C 1989 Edition; Quincy, MA: National Fire Protection Association.

Fire and Explosions in Grain Elevators and Facilities Handling Bulk Raw Agricultural Commodities. NFPA 61B 1989 Edition. Quincy, MA: National Fire Protection Association.

Fire and Explosions in the Milling of Agricultural Commodities for Human Consumption. NFPA 61D 1989 Edition. Quincy, MA: National Fire Protection Association.

Fire and Explosions in Wood Processing and Woodworking Facilities. NFPA 664 1987 Edition. Quincy, MA: National Fire Protection Association.

Fire Hazard Properties of Flammable Liquids, Gases and Volatile Solids. NFPA 325M 1984 Edition. Quincy, MA: National Fire Protection Association.

Haase, H. 1977. *Electrostatic Hazards, Their Evaluation and Control.* New York: Verlag Chemie.

Hazardous Chemicals Data. NFPA 49 1979 Edition. Quincy, MA: National Fire Protection Association.

Nagy, John et al. 1968. Explosibility of miscellaneous dusts. Report of investigations 7208. Washington, DC: U.S. Department of Interior, Bureau of Mines.

National Fuel Gas Code. NFPA 54 1988 Edition. Quincy, MA: National Fire Protection Association.

Spray Application Using Flammable and Combustible Materials. NFPA 34 1987 Edition. Quincy, MA: National Fire Protection Association.

Static Electricity. NFPA 77 1988 Edition. Quincy, MA: National Fire Protection Association.

Storage and Handling of Liquefied Petroleum Gases, NFPA 58 1989 Edition. Quincy, MA: National Fire Protection Association.

10

High Voltage Systems: Design, Construction and Maintenance

Marty Martin

Industrial processes usually require electrical power. This power is used to drive motors, to heat materials, or in electrochemical processes. Often the power requirements of a plant require the electric power to be delivered at high voltage. In this text (and most other codes and texts), high voltage is considered any voltage over 600 V. This voltage could be as high as 138,000 V for some very large facilities. The characteristics of this voltage and the enormous amounts of power being transmitted necessitate special safety considerations.

Safety must be considered during the four activities associated with a high voltage electrical system. These activities are:

1. Design
2. Installation
3. Operation
4. Maintenance

Often the safety considerations are different and unique for each of these activities, but the activities are interrelated. For example, to properly design a safe high voltage system, the designer must be familiar with the installation practices, operation, and maintenance practices of the system. This chapter examines these activities and describes how safety can be enhanced during each of them.

DESIGN

The design of a high voltage system may be as simple as specifying the manufacturer's model number of a unit substation or as complex as designing a 138 kV/12.4 kV overhead substation with a series of 12 kV loops feeding 12.4 kV/480 V double-ended unit substations. However, regardless of the complexity, the designer's responsibility for safety is unique. The designer has the opportunity to minimize the chances of occurrence of many hazardous conditions in the high voltage system, and, indeed, can eliminate many hazardous conditions altogether. By nature, designers thrust their ideas on people who will be very involved in the system. They make decisions about such matters as short-circuit duty and configuration that will greatly affect the people who operate and maintain the system. These people will live (or die!) with the design long after the completion of the design project.

High voltage systems, like many other electrical systems, can grow in hazard as they grow in complexity. The designer of a high voltage system should keep the "KISS" theory in mind—the "Keep It Simple" philosophy. Although there often is a need for complex networking in utility systems, the configuration of industrial systems almost always can be designed in a more simple fashion.

One of the first decisions that a high voltage designer must make concerns the configuration of the high voltage system. The configuration is the arrangement of the high voltage circuit, which can vary according to the reliability requirements of the industrial plant. Industrial high voltage systems usually are of one of the following three types: (1) radial, (2) primary selective, or (3) looped systems.

A radial system consist of a single high voltage source feeding a bus. This bus, in turn, feeds several individual loads. The loads may be transformers, high voltage motors, arc furnaces, or other utilization devices. Where the bus is tapped, an overcurrent protective device is installed. (See Figure 10.1.)

In the radial system, each piece of equipment is fed directly through the source. If the source or the bus were to have an outage, the entire system would be down until the cause of the outage was corrected. This system is very commonly used in industrial application. Its initial cost is low because there is no redundancy; and its inherent safety is good because there is no danger of backfeed, and there is only one source. The system can be very simply drawn and is easy to understand.

Often a designer chooses to introduce a second source for the system. This second source enhances system reliability and also increases the future capacity of the system. The system usually is configured as a primary selective system. This type of system often is used in outdoor industrial substations. (See Figure 10.2.)

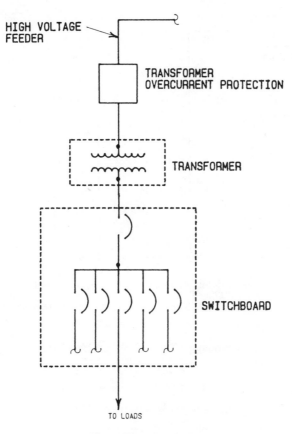

Figure 10.1. Radial system.

In Figure 10.2, the loads are fed from one of two sources. If one source goes down, all loads may be switched to the other source. Although this arrangement enhances reliability, it also causes some safety concerns. Because the complexity of the system has been increased, most designers provide some sort of interlock system so that both switches feeding a piece of equipment may not be closed simultaneously. One can see that if both switches feeding transformer A (A1 and A2) were closed, and the transformer main circuit breaker were opened, the load terminals of feeder circuit breaker No. 2 would be energized. This certainly would introduce a hazardous condition at circuit breaker No. 2, as well as at transformer B. This unintentional paralleling of sources also can introduce hazardous levels of fault current. The interlocks to prevent this may be electrical or mechanical in nature. Typically, a mechanical interlock system is preferred

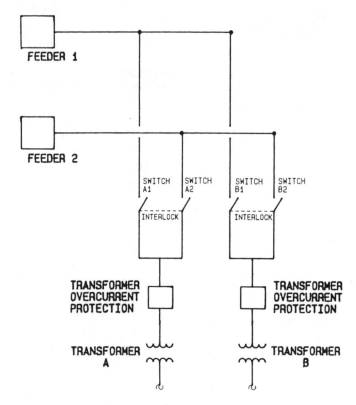

Figure 10.2. Primary selective system.

because it is simple and foolproof. Electrical interlocks are less tamper-resistant and more prone to failure than mechanical interlocks.

A variation on the primary selective system is the looped system. (See Figure 10.3.) This system usually has a normally open switch within the loop. This allows half of the loop to be fed from one source and the other half to be fed from the other source. If a loop cable were to fail, the downstream load would experience an outage. After the damaged piece of cable is isolated and the normally open switch reclosed, the entire load can be reenergized. The design here requires some sort of fault-locating device. These devices, commonly called fault indicators, consist of a winding of wire that goes around the unshielded portion of the high voltage cable or around a high voltage bus. When the fault current passes through this cable, the high fields cause an indicator to flip to a positive position. This position can be easily observed through disconnect windows or upon the opening of an enclosed switchgear door. It tells the operator that the fault was downstream of the indicator. By surveying the indicators, the

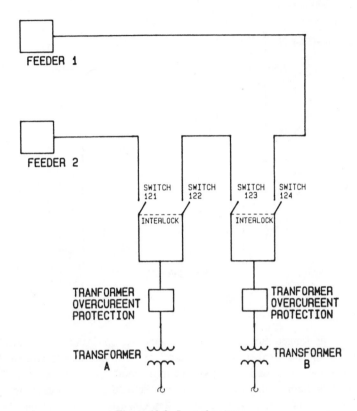

Figure 10.3. Looped system.

operator determines the fault location. This eliminates the need for sectionalizing the loop and trying to reclose, a method that can cause extra damage to faulted equipment and create a hazard to personnel from equipment failure. These fault indicators must be reset.

Often interlocks are placed in the primary loop, isolating switches as well. These interlocks force an operator to have one switch open in the loop at all times. This approach prevents the entire loop from being energized from both sources, with the previously discussed overduty and voltage hazards. It forces a short outage whenever the "normally open" point in the loop is changed.

Another common system configuration of industrial systems is the secondary selective system. (See Figure 10.4.) In a secondary selective system, the low voltage side of the transformer is tied together with a circuit breaker called a tie breaker. When this tie breaker is closed, the entire switchboard is fed from the two transformers. This system allows for greater reliability than the primary selective system because a transformer outage, either planned or unplanned, will

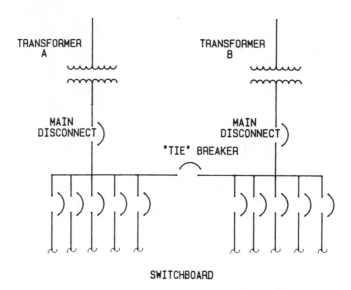

Figure 10.4. Secondary selective system.

not cause a prolonged outage on the secondary. Again, however, the secondary disconnects and tie breaker circuit breakers must be interlocked so that the tie breaker may not be closed while both main disconnects are closed.

After the system configuration is chosen, the actual equipment to be used in the system must be specified. This equipment consists of high voltage switches, circuit breakers, fuses, motor controllers, and metering equipment. The selection of this equipment is a complicated issue, involving decisions that will not be detailed in this text. However, these safety design issues must be considered: (a) fault duty, (2) overcurrent protection, and (3) clearances.

The fault duty for the interrupting rating of a fused disconnect, disconnect, or circuit breaker is the amount of current that a fuse, disconnect, or circuit breaker can interrupt safely. Because these devices will operate when a fault occurs, it is of great importance that they be able to safely interrupt this current. When a switching device is called upon to interrupt a current that exceeds its capacity, several things can happen. In some equipment, such as an air break switch, the arc may not be snuffed out by the equipment. This continuing arc then causes the terminals to melt, showering the area with molten steel. The arc will continue until some upstream overcurrent protective device operates. In other equipment, such as an oil-insulated circuit breaker, the arc will cause instantaneous heating of the vessel. This can cause a dramatic rise of pressure within the vessel, making the vessel rupture. The hazards of these electrically induced explosions cannot be ignored.

In order to calculate the fault current available on the high voltage system, the available fault current must be obtained from the serving utility. If cogeneration is involved in the facility, the contribution of the generator also must be considered. In most industries, motors serve a substantial portion of the electrical load. Because, under faulted conditions, the motors are instantly converted into generators, the contribution of the motor load also must be taken into account. There are several methods and many good computer programs that can assist the designer in the calculation of this available fault current. In large systems, this can be a formidable task. However, Section 110-9 of the National Electrical Code, NFPA 70-1990, states:

Equipment intended to break current at fault levels shall have an interrupting rating sufficient for the system voltage and the current which is available at the line terminals of the equipment.

Equipment intended to break current at other than fault levels shall have an interrupting rating at system voltage sufficient for the current that must be interrupted.

If a facility falls under the jurisdiction of the National Electrical Code or a regional code such as the Chicago Electrical Code, this calculation is mandated by law. A copy of this short-circuit study should be made available to the electrical inspector for review. Not all his voltage equipment interrupts current; equipment such as cable and transformers carries electrical current. During fault conditions, the equipment must carry the high levels of fault current without damage to its insulation strength. It is the heat from this current that damages insulation, and the amount of heat generated is a function of current and time. Overcurrent protection must be selected to interrupt the current before it has time to generate enough heat to damage downstream equipment. Equipment manufacturers give their equipment "withstand ratings" that describe the amount of current (or power) a device can carry for a specified time duration in cycles.

Another facet of the fault current calculation is the bus bracing for various components. During faulted conditions, the magnetic stress on parallel buses is enormous; so mechanical bracing of these buses must be done to account for the available fault current.

Overcurrent protective devices must be specified for all conductors and equipment in a high voltage system. This is required in Article 710 of the National Electric Code. Again, the exact methodology for sizing of overcurrent protective devices will be left to other texts, but it is important to note that overcurrent protection is required to protect all equipment from failure. In low voltage systems, this is a somewhat intuitive process because circuit breakers and fuses on low voltage systems have a nominal current rating. In high voltage systems, however, relay devices and circuit breakers with adjustable trip ratings often are utilized. The adjustability of these devices can complicate the characteristics of the overcurrent protection provided. Withstand ratings and damage curves for cable, transformers, and other non-current-interrupting devices must be consulted in order to establish the relay or circuit breaker settings. These withstand ratings and damage curves are available from the manufacturer's literature.

A third equipment-related design issue is that of clearances. Minimum clearances are detailed in Article 110 of the National Electrical Code (NFPA 70-1990). Before these requirements are described, it must be emphasized that all of these requirements are code minimums. If these dimensions can be improved upon, the safety of those people installing, operating, and maintaining the systems will be enhanced. A clear work space is required around all high voltage electrical equipment. This space should be at least 6′ 6″ high (measured from the floor or platform) and at least 3′ 0″ wide. The depth of the required work space is detailed in Table 110-34(a). These dimensions are measured from live parts that are exposed, or from the enclosure front or opening if they are enclosed. (See Table 10.1.)

The high voltage equipment may be located in an area accessible to qualified people only, or it may be installed in a place accessible to anyone. In modern industrial design, the latter is often the case. The National Electrical Code and good practice dictate that the equipment must be in a secured area or metal-enclosed and labeled with the appropriate caution and high voltage signs. Any doors or access points must be padlocked. Any ventilation openings in a metal enclosure must be baffled or otherwise designed so that foreign objects cannot contact any energized parts. Fenced enclosures must be at least 8 feet high.

If the high voltage equipment is located in an area accessible to qualified persons only, several other design considerations are required. Limiting access to the equipment usually means putting the equipment in a locked room. The size of the room must comply with the aforementioned clearances. Additional working space is highly recommended because of the inflexibility of the walls.

Table 10.1. Minimum Depth of Clear Working Space in Front of Electric Equipment.

	NOMINAL (1)	(2)	CONDITIONS (3)
	(Feet)	(Feet)	(Feet)
601–2,500 V	3	4	5
2,501–9,000 V	4	5	6
9,000–25,000 V	5	6	9
25,000 V–75 kV	6	8	10
Above 75 kV	8	10	12

Any future equipment that is anticipated must be accounted for, and a method of getting the equipment into and out of the room must be determined. The National Electrical Code (Section 110-33) requires at least one entrance not less than 24 inches wide and 6½ feet high, but a wider entrance certainly is recommended. A door that swings out of the vault will allow a more rapid exit during an emergency condition. Having a door at either end of the room also would improve safety.

The high voltage equipment should be well lighted. Illumination should be at least 50 fc. Fixtures should be located so that lamp replacement will not endanger maintenance people. Switch placement should ensure that no one will come in contact with energized parts while turning on the lights. For example, locate the light switch on the outside of the enclosure by the entrance. The lighting should be fed from an emergency generation system, if available. Battery-powered emergency lights are recommended if the room is heated.

Another topic of critical importance to the high voltage system designer is grounding. According to the Institute of Electrical and Electronic Engineers publication *Guide for Safety in A/C Substation Grounding* (ANSI-IEEE Standard 80-1986), grounding design has two objectives:

1. To provide means to carry electric currents into the earth under normal and fault conditions without exceeding any operating and equipment limits or adversely affecting the continuity of service.
2. A person in the vicinity of ground facilities is not exposed to the danger of critical electrical shock.

This guide along with IEEE Standard 142-1982, *Recommended Practice for Grounding Industrial and Commercial Power Systems,* should be referred to in designing the grounding system for an industrial high voltage system. Article 250 of the National Electrical Code also is quite explicit in its requirements for grounding systems.

Several general statements can be made about grounding and high voltage systems. First, all non-current-carrying metallic parts must be grounded and bonded together, to facilitate the operation of overcurrent protective devices and also to minimize the possibility of a voltage potential between two metallic surfaces. It is important to realize however, that during fault conditions there will be differences—indeed, sometime dramatic differences—in potential between grounded surfaces and between different areas of the earth. These differences are due to the large currents that are injected into the earth. Because the earth does have a finite resistance, Ohm's law dictates that when very high current passes through this resistance, a voltage potential must occur. The effect of this potential gradient in the earth can be minimized by the grounding system design.

In a substation, a typical grounding system will include a large conductor (typically 4/0 copper) buried 12 to 18 inches below grade. These conductors are arranged in a grid pattern, with each grid 10 to 20 feet on a side. A similar grid should be installed underneath free-standing metal-enclosed switchgear and all required working space. Ground rods typically are installed at each grid inter-section and at the perimeter of the grid. The total grid resistance of the installed grid should be less than 1 Ohm possible and less than 5 Ohms in all cases. All connections should be extremely heavy duty and meet the IEEE (Institute of Electrical and Electronic Engineers) standard for grounding connections. It is important to note that the areas in which operators will be standing, such as at the operating handles of ground-operated air breaks, at the handles of metal-enclosed gear, or where oil cutouts will be operated, demand special attention. The designer should strive to minimize the voltage between the operator's legs, which are in contact with the earth during equipment operation. (This voltage is often called the "step voltage.") Often, a steel mat is installed at the operating handle. This mat may be on the surface or just below the surface, or may consist of additional ground rods at each operating position.

In outdoor substations an insulating layer of crushed stone will further protect the operator from fault currents, as the current will tend to flow down into the earth and insulate the operator. Another area of special concern is fences. Because the outside of a fence usually is accessible to the public, most engineers agree that fences must be connected to the ground grid system.

The grounding design for portable high voltage equipment is rigorous. Portable high voltage equipment must have a grounded conductor, which must be grounded through an impedance. This impedance limits the amount of current that will flow through a fault in the equipment. Any exposed non-current-carrying metal parts must be connected to an equipment grounding conductor, which is grounded at the same point to which the system neutral (or grounded conductor) impedance is grounded. Ground fault detection is required for the system; it will detect the presence of a ground fault and deenergize the circuit. The voltage developed by this ground fault current may not exceed 100 V. The continuity of the equipment grounding conductor also must be monitored; loss of continuity in the ground conductor must deenergize the circuit. The grounding electrode to which the neutral impedance is connected must be isolated from any other grounds by at least 20 feet.

One of the most important parts of design is documentation. The specifications and drawings for the design must be written in a clear and concise manner. Specifications should be complete, detailing all features of equipment and all pertinent installation details. The drawings should be completed with the installer and user of the system in mind. The installation drawings should show all required clearances. The switching diagram should be clear, with switch numbers and normally opened switches (if any) clearly shown.

The National Electric Code requires that overcurrent protection be provided in each ungrounded conductor by the use of overcurrent relays or fuses. The code recognizes several types of overcurrent protection system.

Circuit breakers for high voltage systems can be installed either indoors or outdoors. Circuit breakers are insulated with mineral oil, sodium hexafluoride, or vacuum. The SF_6 and vacuum-type circuit breakers are most commonly used in new designs today. These circuit breakers must be operated within their continuous current rating, their fault current interrupting rating, their fault closing rating, and their maximum voltage rating. Circuit breakers that are used to control oil-filled transformers located inside a plant must be located outside the transformer vault or be capable of operation from outside the vault, to eliminate the possibility of high fault currents rupturing the tanks of the oil-filled transformers.

High voltage fuses also are used extensively in industrial high voltage systems. The fuses and associated disconnect switches usually are less expensive than circuit breakers. When used as part of metal-enclosed switchgear, they offer an economical alternative to circuit-breaker gear. As with circuit breakers, these fuses must be operated within all continuous current, fault current, and voltage ratings. Some high voltage fuses operate by using a high pressure discharge of gas to "snuff" the arc. When these fuses, often called the solid material discharge type, operate in metal-enclosed switchgear, they offer no hazards as long as the switchgear door is shut. An interlock should be provided to ensure that the doors are shut when the circuit is on. Expulsion-type fuses, which are used in cutouts, often are used in overhead and substation construction. These types of cutouts can be mounted on cross arms or other framing. Older vault designs utilize these fuses; the National Electrical Codes now does not allow that design. These fuses often are closed in one at a time, so that a single phasing condition is put onto three-phase transformers. With sufficient line lengths and small transformer sizes, this could cause a ferroresonant condition. This ferroresonance can cause abnormally high voltage at the transformer, and these voltages can in turn cause flashovers and termination failures.

CONSTRUCTION

After the design has been completed, personnel must construct the high voltage system. The hazards involved in high voltage construction are very similar to the hazards of other types of electrical construction. Some special situations will be encountered, however. The following paragraphs examine both overhead high voltage construction and underground high voltage construction.

Overhead high voltage construction once was very common in industry, but it has for the most part been replaced by underground insulated-conductor type construction. However, there still is a substantial amount of overhead construction. This construction includes open exterior substation and overhead pole lines that serve a campus of several buildings or provide power from the utility point of service to the building. Construction of these systems involves hole or foundation excavation, pouring of the foundation (if any), pole or support framing and erection, conductor installation, and the final connection to the supply system. Guidelines for construction of overhead electrical systems are contained in the Code of Federal Regulations, 29 CFR 1926.950-960. These federal regulations are administered by the Occupational Safety and Health Administration.

The construction of overhead conductor supports must be done with new materials in good condition, used within their recommended mechanical loads. These loads must account for wind loading and, if the geographic areas warrants it, any ice buildup that can occur during ice storms. The maximum mechanical load on a conductor support will be caused by the horizontal wind loading on an ice-covered conductor rather than simply the weight of the conductor. The National Electrical Safety Code (ANSI C2-1990) contains mechanical (and other) requirements concerning this type of construction. The component supplier can provide mechanical strength data to verify the suitability of the construction materials, whether they be wood, steel, aluminum, or insulating material such as ceramics or epoxy type devices.

The erection of poles and other conductor supports presents one of the greatest hazards in high voltage construction: contact with energized lines. Most overhead lines encountered in public and private areas will be bare conductors, that is, noninsulated. Even if these conductors are covered, the covering may not be adequate for the operating voltage of the conductor. For this reason, extreme care must be taken whenever a crane, gin, hoist, or other lifting device is used. If nearby energized lines cannot be deenergized, then insulating barriers, blankets, hoses, or other insulating materials must be installed. The crane operator must examine the site conditions carefully, to locate the crane so that any hazard will be minimized. During erection procedures near energized lines, the area should be roped off to keep the public at distance. Any ground personnel should be instructed to stay away from the crane or line truck during the erection or should wear insulated gloves to protect them from any hazard. Personnel should not stand under poles or supports as they are being erected, and all ropes, cables, slings, and so on, used during the erection must be in good condition and used within their mechanical specifications. If climbing of poles or other structures is necessary during the construction process, it should be done only when poles are backfilled and tamped or support foundations have adequately cured. Wood structures must be inspected to be sure that the pole is sound and that it can take the unbalanced load a climber will put on it. If wire is to be removed or installed, the structure may have to be guyed or shored to accommodate this temporary (or permanent) load.

Before the installation of the conductors, the construction crew should meet to establish guidelines for their installation. In substation work where distances are small, oral communication and established hand signals will suffice. However, in overhead line construction, distances may be so great that radio communication or hand signals must be utilized. Conductors used in overhead line construction should be grounded at the supply reel and at the takeup reel; these grounds allow the cable to be pulled yet maintain a good ground connection. The tensioning reel should be designed to pull the conductors in tension. The supply reels should have an adjustable brake on them to keep the conductor under tension during insulation. The tension and the supply reel should be properly leveled. If any crossings of energized conductors occur, special structures or safety nets should be utilized so that the conductor will not contact the energized line even if it loses tension. Grounds should be installed at either side of the crossing. If the energized line is controlled by an automatically reclosing overcurrent protective device, this reclosing feature should be "locked out" so that if an accidental contact does occur, the device will not reclose onto the fault. Employees should contact the conductors, tensioner, and supply reels only when necessary. High voltage gloves should be worn for any contact. Load ratings must be observed on all pulling ropes, sheaves, and come-alongs, and these devices must be inspected daily. All overhead installation should be stopped if any inclement weather such as rain, high winds, or an electrical storm occurs.

Clearances from overhead high voltage lines to the ground buildings, bridges, swimming pools, and other installations are contained in the National Electrical Safety Code (ANSI C2-1990). This document is published by the Institute of Electrical and Electronic Engineers and applies mainly to utility systems. The National Electrical Code, however, references this document for establishing system clearances in high voltage systems. It should be noted that these clearances are minimums. Because the amount that a conductor will sag is dependent upon temperature, these minimum clearances must be adjusted for construction.

Underground high voltage construction consists of direct buried cables, cables pulled into conduit systems, pad-mounted equipment, and the construction of concrete-encased ductwork/manhole systems. Construction hazards and safety practices are similar to those of other excavation and construction situations.

The first thing one must do when any underground construction is planned is to call all local utilities to determine the location of existing underground electrical, telephone, television, gas, steam, water, and sewer utility locations. If any existing site plans are available, they may provide clues to the location of old foundations, underground tank systems, or other underground obstructions.

Safety regulations for excavation are contained in the Code of Federal Regulations, 29 CFR 1926.650. All of these regulations are important, but some have special meaning for high voltage construction. Once the location of all existing utilities is known, excavation can begin. If any existing utilities are unearthed during excavation, they must be supported and guarded. The excavation area itself must be guarded from public entry. Signage and warning lights must be included. If the excavation area is exposed to vehicular traffic, the construction crews must wear high-visibility clothing. The public and the employees must be kept away from any mechanical digging, trenching, or loading or unloading of equipment. If a trench is more than 5 feet deep, OSHA requires that sheeting or sheet piles must be installed to shore up the sides of the excavation. In certain cases, the sides of the trench may be angled to eliminate the need to sheet piling or shoring. The angle will depend upon the field conditions; the trench depth, existing material type, water content, any loading imposed by adjacent structures, any changes to materials such as sun, water, or freezing, and the effect of vibration by equipment or blasting will all affect the excavation angle. The Code of Federal Regulations should be consulted. Excavation material should be kept at least 2 feet and preferably farther away from the trench. If any above-grade structures will be approached by the excavation, shoring or underpinning may be required. If the trench is deeper than 4 feet, a ladder should be provided every 25 feet to facilitate the rapid evacuation of the trench. Any trenching or manhole excavation should be inspected daily and after any precipitation to determine the soil stability. If the soil stability is in doubt, shoring should be done before any personnel enter the excavated area.

Safety requirements for lifting precast manholes, pad-mounted switchgear, or other equipment are detailed in the Code of Federal Regulations, 29 CFR 1926-50. Among the pertinent rules are requirements for operating items of equipment below their mechanical capacities and inspection requirements for lifting cables, ropes, and slings.

The installation of cables in conduit systems involves the insertion of ropes into the systems and the pulling of the cables through the conduit systems with the ropes. Again, all ropes, rope pulling equipment, and any sheaves, pulleys, or supports used in the process must be used within their rated capacities. Workers should stand away from the direct path of a cable in case something breaks and causes a backlash. If a capstan type of pulling device is used, care must be taken that loose clothing or hands do not get caught in the capstan. The manufacturer's recommendations for using these cable pullers must be observed.

An important yet often ignored part of the installation process is the correction and/or updating of the design drawings. The installer should have the design drawings corrected to reflect what actually was installed. Often the location of underground conflicts as well as any changes in the switching or cable routing can be noted on detailed drawings or on the one-line diagram. This record drawing set will be invaluable to the people who will operate and maintain the system.

SYSTEM OPERATION AND MAINTENANCE

The third and fourth facets of high voltage safety may be discussed together. In high voltage systems, maintenance should be a part of system operation. Preventive and/or predictive maintenance is required so that the reliability of the high voltage system can be maintained.

Before any discussion of the pertinent details of safety in the operation and maintenance of high voltage systems, the critical need for first aid training must be emphasized. All employees who will work with or on high voltage electrical systems must have training in first aid, including cardiopulmonary resuscitation techniques, emergency breathing techniques, and first aid treatment for electrical shock, burns, falls, or other trauma associated with electricity. The Red Cross can be contacted for information regarding courses in first aid. These skills must be continually reinforced with constant retraining and updating. Retraining every six months is recommended for people working with electrical systems.

Initially, much electrical safety equipment must be purchased after the construction of a high voltage system. The function of most of these tools is to make Ohm's law work in the user's favor, that is, to increase the resistance between the hazardous voltage and the worker, so that no (or an extremely small amount of) current will flow. These tools include personal insulating equipment such as rubber gloves and sleeves, fire suits, eye shields, and hard hats. This equipment must be used according to the manufacturer's recommendations. For example, gloves should be retested monthly, and an air test should be performed on the insulating gloves every time they are put on. This air test consists of rolling up the open end of the glove, trapping air inside, and then inspecting the glove for air leaks. Rubber blankets and hoses also may be required; this equipment is placed on energized equipment to protect workers from accidental contact. The blankets and the hoses also must be periodically tested.

Fiberglass sticks, commonly called "hot sticks," are used for many switching and maintenance operations on high voltage systems. The purpose of the sticks is to keep personnel a safe distance away from equipment when it is energized or while it is being energized. The sticks typically are made of fiberglass and usually are rated at 100,000 V per foot for 5 minutes. The correct type of sticks must be procured for each installation. Some sticks have universal equipment mounting heads on them, whereas others are good for a single purpose such as current readings, voltage testing, or fuse removal. These sticks must be periodically cleaned and must be stored in a clean, dry environment. They should not be allowed to contact the ground. Safe ladders also must be available. These ladders should be nonconducting and long enough to provide access to all required equipment.

Much work on substations and overhead equipment is done from elevated work platforms. These platforms may be trailer-mounted or mounted on a line truck. The insulating quality of the buckets and the arm must be periodically tested. Any hydraulic or pneumatic tools on the platform must have nonconducting hoses. When these elevated work platforms are in use, the area around the truck must be secured from the public, and other workers should consider the platform to be energized if they are working near energized lines. The insulating qualities of this equipment must be tested continually.

In underground construction, manhole ventilation systems and water pumps are required to make the manhole for vaults safe for entry. Safety regulations for work in underground manholes are contained in the Code of Federal Regulations, 29 CFR 1926.956.

Maintenance of high voltage equipment should be done with the equipment deenergized whenever possible. Although there may be some inconvenience because of the outages, or workers may have to come in after hours to perform this maintenance, the safety value of this approach cannot be overestimated. However, even when one is working with deenergized equipment, the procedure for deenergizing and reenergizing the equipment must be firmly established and strictly adhered to.

Before any planned maintenance of high voltage equipment has begun, all parties involved in the procedure must meet and discuss all facets of the procedure. Each participant must understand the operation and the sequence of events for maintenance. Communications before and during the maintenance procedure are vital for safety. During the procedure, when a member of the maintenance team reports a change in status of the system or describes an operation that he or she is performing, the other person or persons involved with the procedure should repeat the communications aloud. This is done to ensure that there is no misunderstanding, and it helps to maintain the maintenance team's mental alertness.

All electrical equipment should be considered energized until it has been isolated and grounded. After a procedure has been established regarding which switch or switches are to be opened to deenergize the system in question, the lead person in the maintenance team will open the switches. Inspection of the design drawings and physical inspection of the actual installation must be done to ensure that the switches are the correct ones and the only switches required. After opening of the disconnect, a padlockable hold card should be installed in the disconnect handle. This locking device ensures that the switch will not be closed during the operation. Each employee who will be contacting the energized equipment should place his or her personal padlock on the locking device; this ensures that each member of the maintenance team will acknowledge the reenergization of the system. Use of a simple hold card stating that the system must not be turned back on simply is not enough. After the switch has been opened, any controls such as electrical controls or stored energy devices such as springs should be disconnected or discharged to eliminate the possibility of reenergization. These devices too should be locked out by the above-described padlock scheme. After the device has been disconnected from the source, a special hot stick, called a "fuzz stick," should be used to determine if any voltage exists on the isolated equipment. This fuzz stick uses a capacitive discharge system to detect voltage. If part of the system in question contains capacitors or long lengths of shielded cable, a voltage may remain on the system after deenergization. With capacitors, a 30-minute period after the deenergization should be observed before any other action is taken. After a negative reading with the fuzz stick, grounds should be installed on the system. These grounds should be at the area of work and if available on either side (electrically) of the work area. These grounds should be installed with a hot stick connecting the grounded end first and the equipment end second. The electrical device then may be considered deenergized. If any energized parts are near the work area, they must be isolated by shields or rubber blankets.

After the work has been completed, all members of the maintenance team should verify that his or her work has been done. After all employees have reported clear, the team leader shall remove all protective grounds. The team members then remove their padlocks from the holding device, and the team leader shall perform a visual inspection of the system. At that time, the switches may be closed.

If the high voltage system may not be deenergized, then the extreme hazard of working on energized high voltage equipment must be recognized and analyzed by careful planning.

Authority for live line work must be granted by company management; it is

Table 10.2. Alternating Current Minimum Distances.

VOLTAGE RANGE (PHASE TO PHASE), KILOVOLTS	MINIMUM WORKING AND CLEAR HOT STICK DISTANCE
2.1 to 15	2'0"
15.1 to 35	2'4"
35.1 to 46	2'6"
46.1 to 72.5	3'0"
72.6 to 121	3'4"
138 to 145	3'6"
161 to 169	3'8"
230 to 242	5'0"
345 to 362	7'0"
500 to 552	11'0"
700 to 765	15'0"

Note. For 345 to 362 kV, 500 to 552 kV, and 700 to 765 kV, the minimum working distance and the minimum clear hot stick distance may be reduced, provided that such distances are not less than the storage distance between the energized part and the grounded device.

important to have management personnel involved in a decision not to deenergize equipment. All live line work should be done by people who have been trained in the use of all required tools and specifically in hot line work. It is suggested that a written procedure for each live line operation be developed and agreed to by all maintenance team members. After the plan has been established and agreed upon, all required tools and safety equipment should be assembled. The area of the work should be corded off so that uninvolved parties may not get within 10 feet of the actual work. It is important that all safety equipment, such as gloves, glove protectors, sleeves, hard hats, face shields and protective suits, be utilized.

OSHA regulations dictate a minimum working distance for persons working on energized equipment. No one shall be permitted to approach or to take any conductive object without an approved insulated handle near to energized parts. Safe working clearances for various voltage levels are presented in Table 10.2.

The minimum clearances shown in the table should not be penetrated by the maintenance team members with any body part. During the live line operation, any member of the team working within the minimum distance should communicate orally with the other members of the team, describing the action that is taken, and the members of the team should keep the worker advised of any outside action. The work area should be well illuminated and dry.

REFERENCES

National Electrical Code. NFPA 70—1990 Edition, Quincy, MA: National Fire Protection Association.

National Electrical Safety Code. ANSI C2-1990. New York: Institute of Electrical and Electronic Engineers.

IEEE Standard 80-1986. *A Guide for Safety in AC Substation Grounding.* New York: Institute of Electrical and Electronic Engineers.

IEEE Standard 142-1982. *Recommended Practice for Grounding for Industrial and Commercial Power Systems.* New York: Institute of Electrical and Electronic Engineers.

Index

Index